Marine Corps Martial Arts Program (MCMAP)

US Marine Corps

DISTRIBUTION STATEMENT B: Distribution authorized to U.S. Government agencies for official use only.

PCN 144 000066 00

DEPARTMENT OF THE NAVY
HEADQUARTERS UNITED STATES MARINE CORPS
WASHINGTON, D.C. 20380-1775

FOREWORD

21 Nov 2011

Marine Corps Reference Publication (MCRP) 3-02B, *Marine Corps Martial Arts Program (MCMAP)*, is designed for Marines to review and study techniques after receiving initial training from a certified Marine Corps martial arts instructor or martial arts instructor trainer. It is not designed as a self-study or independent course. The true value of Marine Corps Martial Arts Program is enhancement to unit training. A fully implemented program can help instill unit esprit de corps and help foster the mental, character, and physical development of the individual Marine in the unit. This publication guides individual Marines, unit leaders, and martial arts instructors/instructor trainers in the proper tactics, techniques, and procedures for martial arts training. MCRP 3-02B is not intended to replace supervision by appropriate unit leaders and martial arts instruction by qualified instructors. Its role is to ensure standardized execution of tactics, techniques, and procedures throughout the Marine Corps. Although not directive, this publication is intended for use as a reference by all Marines in developing individual and unit martial arts programs. For policy on conducting martial arts training, refer to Marine Corps Order 1500.59, *Marine Corps Martial Arts Program (MCMAP)*.

This publication supersedes MCRP 3-02B, *Marine Corps Martial Arts* (formerly *Close Combat*), dated 18 February 1999.

WARNING

Techniques described in this manual can cause serious injury or death. Practical application in the training of these techniques will be conducted in strict adherence with training procedures outlined in this manual as well as by conducting a thorough operational risk assessment for all training.

Reviewed and approved this date.

BY DIRECTION OF THE COMMANDANT OF THE MARINE CORPS

RICHARD P. MILLS
Lieutenant General, U.S. Marine Corps
Deputy Commandant for Combat Development and Integration

Publication Control Number: 144 000066 00

DISTRIBUTION STATEMENT B: Distribution authorized to U.S. Government agencies only; for official use only. Other request for this document will be referred to Headquarters United States Marine Corps, Combat Development and Integration, Capabilities Development Directorate, Fires and Maneuver Integration Division, Quantico, Va.

For Official Use Only

THIS PAGE INTENTIONALLY LEFT BLANK.

Marine Corps Martial Arts Program

TABLE OF CONTENTS

Chapter 1. Overview

Structure	1-2
Belt Ranking System	1-3
Testing	1-4
Martial Arts Instructor	1-4
Martial Arts Instructor Trainer	1-5
Tactics and Techniques	1-5

Chapter 2. Tan Belt

Section I: Fundamentals of the Marine Corps Martial Arts Program	2-2
Ranges of Close Combat Engagements	2-2
Target Areas of the Body	2-2
Weapons of the Body	2-3
Basic Warrior Stance	2-3
Angles of Movement	2-4
Section II: Punches	2-6
Muscle Relaxation	2-6
Making a Fist	2-6
Weight Transfer	2-7
Rapid Retraction	2-7
Telegraphing	2-8
Lead Hand Punch	2-8
Rear Hand Punch	2-9
Uppercut	2-10
Hook	2-11
Actions of the Aggressor	2-13
Section III: Break-Falls	2-15
Front Break-Fall	2-15
Back Break-Fall	2-16
Side Break-Fall	2-17
Forward Shoulder Roll	2-19

For Official Use Only

Section IV: Bayonet Techniques 2-21
    Straight Thrust 2-21
    Horizontal Buttstroke 2-22
    Vertical Buttstroke 2-23
    Smash 2-24
    Slash 2-25
    Disrupt 2-26

Section V: Upper Body Strikes 2-28
    Principles 2-28
    Hammer Fist 2-29
    Eye Gouge 2-31
    Vertical Elbow Strike (Low to High) 2-32
    Forward Horizontal Elbow Strike 2-33
    Actions of the Aggressor 2-34

Section VI: Lower Body Strikes 2-36
    Vertical Knee Strike 2-36
    Front Kick 2-36
    Round Kick 2-37
    Vertical Stomp 2-38
    Actions of the Aggressor 2-40

Section VII: Introduction to Chokes 2-41
    Rear Choke 2-41
    Figure Four Variation 2-43

Section VIII: Throws 2-45
    Throw 2-45
    Actions of the Aggressor 2-47

Section IX: Counters to Strikes 2-49
    Counter to Strikes 2-49
    Counter to Rear Hand Punch 2-49
    Counter to the Rear Leg Kick 2-50

Section X: Counters to Chokes and Holds 2-52
    Counter to the Rear Choke 2-52
    Counter to the Rear Headlock 2-53
    Counter to the Rear Bear Hug 2-54

Section XI: Unarmed Manipulation 2-56
    Compliance Techniques 2-56
    Distracter Techniques 2-56

For Official Use Only

Joint Manipulation 2-57
Wristlocks 2-57
Armbar Takedown 2-58

Section XII: Armed Manipulations 2-62
Rifle and Shotgun Retention Techniques 2-62
Blocks 2-65

Section XIII: Knife Fighting 2-69
Principles of Knife Fighting 2-69
Vertical Slash 2-71
Vertical Thrust 2-72

Chapter 3. Gray Belt

Section I: Bayonet 3-2
Execution 3-2
Movement 3-2
Closing 3-3
Bayonet Techniques Disrupt and Thrust While Closing 3-3

Section II: Upper Body Strikes 3-6
Chin Jab/Palm Heel Strike 3-6
Knife Hand Strikes 3-7
Inside/Reverse Knife Hand Strike 3-8
Vertical Knife Hand Strike 3-9
Elbow Strikes 3-10
The Rear Horizontal Elbow Strike 3-11
Vertical Elbow Strike (High to Low) 3-11
Actions of the Aggressor 3-12

Section III: Lower Body Strikes 3-14
Horizontal Knee Strike 3-14
Side Kick 3-14
Axe Stomp 3-16
Actions of the Aggressor 3-17

Section IV: Front Choke 3-18

Section V: Hip Throw 3-20

Section VI: Counters to Strikes 3-22
   Counter to the Lead Hand Punch 3-22
   Counter to a Lead Leg Kick 3-24

Section VII: Counters to Chokes and Holds 3-25
   Counter to the Front Choke 3-25
   Counter to the Front Headlock 3-26
   Counter to the Front Bear Hug 3-28

Section VIII: Unarmed Manipulation 3-30
   Wristlock Come-Along 3-30
   Takedown From a Wristlock Come-Along and Double Flexi Cuff 3-32
   Escort Position 3-35
   Escort Position Takedown and Single Flexi Cuff 3-36

Section IX: Armed Manipulation 3-38
   Aggressor Grabs With Both Hands (Pushing) 3-38
   Aggressor Grabs With Both Hands Pulling (Stationary) 3-40
   Aggressor Grabs With Both Hands Pulling (Moving) 3-42
   Aggressor Grabs Over Handed With Strikes 3-42
   Aggressor Grabs Under Handed With Strikes 3-45

Section X: Knife Techniques 3-47
   Forward Thrust 3-47
   Forward Slash 3-48
   Reverse Thrust 3-49
   Reverse Slash 3-51
   Bulldogging 3-52

Section XI: Weapons of Opportunity 3-54
   Straight Thrust 3-54
   Vertical Strike 3-55
   Forward Strike 3-56
   Reverse Strike 3-57

Section XII: Ground Fighting 3-59
   Counter to Mount Position 3-59
   Counter to Guard Position 3-60

Chapter 4. Green Belt

Section I: Bayonet Techniques 4-2
   Fundamentals 4-2

Execute a Disrupt and a Thrust While Closing With a Moving Aggressor 4-4
Buttstroke Offline 4-4

Section II: Side Choke 4-7

Section III: Shoulder Throw 4-9

Section IV: Counters to Strikes 4-11
Counter to a Round Punch 4-11
Counter to Round Kick 4-13

Section V. Lower Body Strikes 4-15

Section VI: Unarmed Manipulations 4-17
Enhanced Pain Compliance 4-17
Reverse Wristlock Come-Along 4-20
Controlling Techniques 4-22

Section VII: Knife Techniques 4-25
Counter to a Vertical Strike With Follow-on Techniques 4-25
Counter to a Forward Strike With Follow-on Techniques 4-26

Section VIII: Weapons of Opportunity 4-27
Block for a Vertical Strike With Follow-on Strikes 4-27
Block for a Forward Strike With Follow-on Strikes 4-28
Block for a Reverse Strike With Follow-on Strikes 4-29
Block for a Straight Thrust With Follow-on Strikes 4-30

Section IX: Ground Fighting 4-32
Armbar From Mount Position 4-32
Armbar From Guard Position 4-34

Chapter 5. Brown Belt

Section I: Bayonet Techniques 5-2
One-on-Two Engagement 5-2
Two-on-One Engagement 5-3
Two-on-Two Engagement 5-4

Section II: Ground Fighting 5-6
Ground Fighting Techniques 5-6
Basic Leg Lock 5-7

For Official Use Only

Section III: Ground Chokes 5-9
Rear Ground Choke 5-9
Figure-4 Variation of Rear Choke 5-10
Front Choke 5-11
Side Choke 5-12

Section IV: Major Outside Reap Throw 5-14
Major Outside Reap Throw: Aggressor Pushing 5-14
Major Outside Reap Throw: Aggressor Pulling 5-16

Section V: Unarmed Versus Handheld 5-18
Hollowing Out with Follow-on Technique 5-18
Forward Armbar Counter 5-19
Reverse Armbar Counter 5-20
Bent Armbar Counter 5-22

Section VI: Firearm Retention 5-24
Blocking Technique 5-24
Armbar Technique 5-25
Wristlock Technique 5-25
Same Side Grab: From Front 5-27
Same Side Grab: From Rear 5-29

Section VII: Firearm Disarmament 5-31
Counter to Pistol to: Front 5-31
Counter to Pistol to: Rear 5-31

Section VIII: Knife Techniques 5-35
Block for a Reverse Strike 5-35
Block for a Straight Thrust 5-36

Chapter 6. Black Belt

Section I: Bayonet Techniques 6-2
Bayonet Training: Stage One 6-2
Bayonet Training: Stage Two 6-2
Bayonet Training: Stage Three 6-2
Common Error 6-3

Section II: Sweeping Hip Throw 6-4
Section III: Ground Fighting 6-6
Face Rip From the Guard.. 6-6
Straight Armbar From a Scarf Hold.. 6-8
Bent Armbar From a Scarf Hold... 6-10
Section IV: Unarmed Manipulation: Neck-Crank Takedown 6-11
Section V: Chokes 6-13
Triangle Choke.. 6-13
Guillotine Choke 6-15
Section VI: Knee Bar 6-16
Rolling Knee Bar 6-16
Sitting Knee Bar.. 6-18
Section VII: Counter to Pistol to the Head 6-20
Counter to Pistol to the Head: Two Handed 6-20
Counter to Pistil to the Head: One Handed.. 6-22
Section VIII: Upper Body Strikes 6-24
Cupped Hand Strike.. 6-24
Face Smash 6-24
Section IX: Knife Techniques 6-26
Lead Hand Knife... 6-26
Slashing Techniques 6-26
Thrusting Techniques.. 6-28
Reverse-Grip Knife Techniques 6-31
Section X: Improvised Weapons... 6-36
Garrote From the Rear 6-36
Garrote From the Front 6-39

Appendices

A   Training Safety Sheet

Glossary

References

For Official Use Only

THIS PAGE INTENTIONALLY LEFT BLANK.

The techniques are written for right-handed execution. To train left-handed, reverse the lefts and rights in the descriptions. To be truly proficient at all techniques, develop each technique both right-handed and left-handed (strong side/weak side) during sustainment training. Each technique has its general applicable safeties numbered in its introduction. The numbers correspond to the safeties listed in appendix A.

THIS PAGE INTENTIONALLY LEFT BLANK.

CHAPTER 1   *Overview*

The focus of Marine Corps Martial Arts Program (MCMAP) is the personal development of each Marine in a team framework using a standardized, trainable, and sustainable close combat fighting system. As a weapon-based system, all techniques are integrated with equipment, physical challenges, and tactics found on the modern battlefield. The MCMAP is designed to increase the warfighting capabilities of individual Marines and units, enhance Marines' self-confidence and esprit de corps, and foster the warrior ethos in all Marines. The MCMAP is a weapon-based system rooted in the credo that every Marine is a rifleman and will engage the aggressor from 500 meters to close quarter combat. The MCMAP:

- Enhances the Marine Corps' capabilities as an elite fighting force.
- Provides basic combative skills for all Marines.
- Applies across the spectrum of violence.
- Strengthens the Marine Corps warrior ethos.

The motto of MCMAP best states the essence of the program: "One mind, any weapon." This means that every Marine is always armed even without a weapon. He is armed with a combat mindset, the ability to assess and to act, and the knowledge that all Marines can rely on one another.

The Marine Corps was born during the battles that created this country. Drawing upon the experiences of the first Marines, we have developed a martial culture unrivaled in the world today. This legacy includes not only our fighting prowess but also the character and soul of what makes us unique as Marines. This training continued to evolve up to World War II. During these early years, the leadership and core values training that are our hallmark today developed in concert with the martial skills.

As World War II burst upon the Corps, individuals and units were developing specialized training based upon experiences of Marines from the interwar years. This included exposure of Marines to far eastern martial arts systems such as judo and karate. These various systems such as combat hitting skills, the O'Niel System, and those of the Marine Raiders were employed by Marines during the island-hopping campaigns. Additionally, the rapid expansion of the Marine Corps saw a refinement of character and leadership development programs. This trend continued after World War II through the post Vietnam war period. Marines tested and refined new techniques adapted to the technologies and the innovation of a new generation of Marines. In response to societal changes after Vietnam, professional military education and structured leadership training became the focus during the 1970s and 1980s.

For Official Use Only

The 1980s saw the development of the linear infighting neural-override engagement (LINE) system. The LINE system, developed in response to a perceived need for a standardized close combat system, was an important step in the evolution of a Marine Corps specific martial arts program. In 1996 a review of the LINE system, combat hitting skills, pugil stick training, and lessons learned from past programs were combined with the input of 10 subject matter experts from numerous martial arts disciplines to develop the Marine Corps Close Combat Program. This program combined all aspects of close combat into one program. In June 1999, Commandant James L. Jones detailed his vision of a martial arts program and, with his guidance, the close combat program underwent a period of testing and review that resulted in MCMAP. The MCMAP evolved into its present day form by combining the best combat-tested martial arts skills and time-honored close combat training techniques with proven Marine Corps core values and leadership training.

Structure

The MCMAP consists of three disciplines: mental, character, and physical. Each discipline is divided and presented systematically to Marines at each belt level. Those disciplines taught at lower belt levels are then reviewed and reinforced during follow-on training and at the next belt level. Many skills specific to one discipline reinforce the strengths of the other disciplines. Martial culture studies strengthen the mental character of Marines through the historical study of war, at the same time reinforcing the importance of character to a warrior and a martial society.

What makes this a complete program is the synergy of all three disciplines, which are inextricably linked to each other, and to the advancement process within the belt ranking system. Commanders are required to certify that the Marine meets annual training requirements and the prerequisites of each specific belt level and possesses the maturity, judgment, and moral character required for advancement. This ensures that as a Marine develops increased levels of lethality with additional physical skills, he also develops a commensurate level of maturity and self-discipline.

Mental Discipline

Mental discipline has two main components, warfighting and PME, that encompasses the following:

- Warfighting:
    - The study of the art of war.
    - Tactical decisionmaking training.
    - The tactics, techniques, and procedures of expeditionary maneuver warfare.
    - Marine Corps common skills training.
    - Force protection and operational risk management.
- PME:
    - The professional reading program.
    - Martial culture studies.
    - The historical study of war.
    - The study of Marine Corps history, customs, courtesies, and traditions.

Mental discipline creates a smarter Marine, who is capable of understanding and handling the complexity of modern warfare—a Marine who is tactically and technically proficient. This training begins with the transformation at recruit training and serves as the foundation of the future leadership of the Corps.

Character Discipline

Character discipline encompasses the study of the human dimensions of combat and has two main components: the Marine Corps core values program; which consists of honor, courage, and commitment, and the Marine Corps leadership training program; which consists of mentoring, and leadership schools and courses. This discipline is designed to instill the Marine Corps ethos into every Marine. Character discipline is the spiritual aspect of each Marine and the collective spirit of the Marine Corps. The components of the character discipline instill the warrior spirit in every Marine and emphasize the best traditions for developing esprit de corps, camaraderie, and a warrior mindset. By building the character of a Marine, the Marine develops as a warrior-defender who embodies the core values and who is self-disciplined, confident, and capable of making the right decision under any condition, from combat to liberty. The proverbial ethical warrior.

Physical Discipline

Physical discipline consists of two main components: fighting techniques and the combat conditioning program. The fighting techniques are trained as part of MCMAP, a weapons-based system beginning at assault fire and moving to the four elements of the fighting component (rifle and bayonet, edged weapons, weapons of opportunity, and unarmed combat). All four elements have a role across the entire spectrum of combat. The various armed and unarmed combat techniques are combined with physical preparedness into the Marine Corps combat conditioning program. It is the sinew of the Corps and prepares every Marine for his role as a basic rifleman—to seek out, close with, and destroy the enemy by fire and movement or repel his assault by fire and close combat. The Marine Corps combat conditioning program:

- Is battlefield oriented.
- Is based on being equipped for combat and develops the Marine's ability to overcome physical hardship and physical obstacles under any climatic condition and in any geographic location.
- Is designed on the philosophy that there are no time limits, level playing fields, or second chances in combat.
- Develops a physical toughness in every Marine that will translate into mental toughness.
- Produces Marines who possess combat fitness and the ability to handle any situation that confronts them.

Belt Ranking System

The MCMAP consists of a belt ranking system with five basic levels: tan, gray, green, brown, and black belt (see table 1-1 on page 1-4). The colored belt levels are identified as user levels and are designed to progressively develop the physical skills and lethality of the individual Marine so that he becomes a stronger link in the Marine Corps chain. The user's responsibilities include participating in all technique classes, tie-ins, warrior studies, and sustaining techniques. Marines must also participate in the appropriate belt-level drills and free sparring. Marines are responsible at all times for their individual actions and conduct.

Table 1-1 Marine Corps Martial Arts Program Belt Ranking System.

Tan Belt	Conducted at entry-level training as part of the transformation process.
	Requires proficiency in basic techniques and a basic understanding of Marine Corps leadership and core values.
Gray Belt	Follow-on training after entry level.
	Builds on the basics with introduction to intermediate techniques.
	Requires mastery of Tan Belt techniques and proficiency of Gray Belt techniques along with continued mental and character discipline training.
	Qualified to attend the instructor course and become MOS 0916.
Green Belt	Skill level for the noncommissioned officers.
	Continued development of intermediate level training.
	Requires mastery of Tan through Gray Belt techniques and proficiency of Green Belt techniques.
	Leadership and core values development training and PME requirements.
Brown Belt	Continued intermediate level training as well as introduction to advance techniques.
	Mastery of Tan through Green Belt techniques and proficiency of Brown Belt techniques. D
	Develop ability to teach leadership and core values training.
Black Belt: 1st Degree	Becoming a serious student of the martial arts.
	Advanced level skills training begins in earnest.
	Mastery of Tan through Brown Belt techniques and proficiency in Black Belt techniques.
	Proven leader and mentor.
Black Belt: 2d degree through 6th degree	Continued development and mastery of all components of the various disciplines.
	Proven leader, teacher, and mentor.

Testing

Advancement in the belt ranking system includes meeting mental and character discipline requirements and the prerequisites for each belt level. Each Marine will be required to show that he has maintained proficiency in the physical disciplines of his current belt and the physical disciplines of the next belt level.

Martial Arts Instructor

The martial arts instructor develops users at the small unit level and is responsible for teaching up to their own belt certification. An instructor teaches the physical techniques that are the building blocks of our physical discipline and develops a unit's character and mental training to positively influence cohesion, esprit de corps, and readiness.

Martial Arts Instructor Trainer

The martial arts instructor trainer (MAIT) develops instructors within his unit, develops a unit level plan, and coordinates the resources and abilities of the instructors within the unit. The MAIT course is a 7 week, military occupational specialty (MOS) 0917-producing course held at the lead school in Quantico, VA. A MAIT can run the instructor courses at the unit level and also train and test Marines up to black belt instructor. Course attendance requirements are Green Belt instructor or higher, sergeant or above with appropriate level PME completed, commander's consent, a first class physical fitness test, and medical screening.

Tactics and Techniques

The MCMAP provides the individual Marine with a set of weapons-based and unarmed techniques for engaging in close combat across the spectrum of violence. Individual techniques can be combined into various procedures to fit the tactical situation. The MCMAP techniques, in the following subparagraphs, are key Marine Corps tactical concepts that are not standalone ideas but are to be combined to achieve an effect that is greater than their separate sum.

Achieving a Decision

An indecisive fight creates a loss of energy and possibly Marines' lives. Whether the intent is to control an aggressor through restraint or defend themselves in war, Marines must have a clear purpose before engaging in close combat and act decisively once engaged.

Gaining an Advantage

A basic principle of martial arts is to use the aggressor's strength and momentum against him to gain more leverage than one's own muscles alone can generate, thereby gaining an advantage. Achieving surprise through deception or stealth can also greatly increase leverage. In close combat, Marines must exploit every advantage over an aggressor to ensure a successful outcome. This can include employing various weapons and close combat techniques that will present a dilemma to the aggressor.

Speed

In close combat, the speed and violence of the attack against an aggressor provides Marines with a distinct advantage. Marines must know and understand the basics of close combat so that they can act instinctively with speed to execute close combat techniques.

Adapting

Close combat can be characterized by chaos, friction, uncertainty, disorder, and rapid change. Each situation is a unique combination of shifting factors that cannot be controlled with precision or certainty. For example, a crowd control mission may call for Marines to employ various techniques ranging from nonlethal restraint to more forceful applications. Marines who adapt quickly will have a significant advantage.

Exploiting Success

An aggressor will not typically surrender simply because he was placed at a disadvantage. Marines cannot be satisfied with gaining an advantage in a close combat situation. Marines must exploit any advantage aggressively and forcefully until an opportunity arises to completely dominate the aggressor. Marines must exploit success by using every advantage that can be gained.

CHAPTER 2: Tan Belt

The Tan Belt is the first belt level in MCMAP and is the minimum basic requirement for all Marines. Tan Belt training includes the introduction to the basic fundamentals of the mental, character, and physical disciplines of MCMAP conducted during entry-level training. Warfighting concepts, character values, and the basic fighting techniques that are required of a basically trained Marine are also included in Tan Belt training. In addition to MCMAP requirements, Tan Belt training and readiness events are also a component of Marine Corps common skills volume I in accordance with Marine Corps Order P3500.72A, *Marine Corps Ground Training and Readiness (T & R) Program*.

Tan Belt Requirements.

Prerequisites	None
Training Hours	Minimum of 27.5 hours of instruction, excluding remedial practice time and testing
Sustainment Hours	None

For Official Use Only

Section I
Fundamentals of the Marine Corps Martial Arts Program

The MCMAP trains Marines how to execute unarmed and armed techniques, in close proximity to another individual, across a spectrum of violence within a continuum of force. Unarmed techniques include hand-to-hand combat and defense against handheld weapons. Armed techniques include techniques applied with table of organization weapons, edged weapons or weapons of opportunity.

Ranges of Close Combat Engagements

There are three ranges in which close combat engagements can take place: long-range, midrange, and close range. In any engagement, these ranges can rapidly transition from one to another and then back again until the aggressor is defeated or the situation is handled. At long-range, the distance between combatants allows engagement with a rifle and bayonet or weapons of opportunity. At midrange, the distance between combatants is such that they can engage each other with knives, punches, or kicks. At close range, the distance between combatants is such that they can grab a hold of each other and may involve elbow and knee strikes and grappling.

Target Areas of the Body

During close combat engagements, the parts of the aggressor's body that are readily accessible will vary with each situation and throughout the confrontation. The goal is to attack the areas that are readily accessible. These target areas are the head, neck, torso, groin and extremities. See figure 2-1.

Head

The vulnerable regions of the head are the eyes, temple, nose, ears, and jaw. Massive damage to the skull can kill an aggressor.

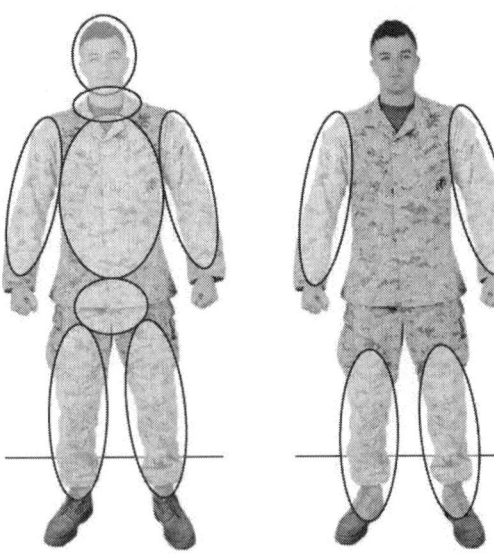

Target Areas   Weapons of the Body

Figure 2-1. Target Areas and Weapons of the Body.

Neck

The entire neck is vulnerable as it contains vital blood vessels, the trachea, and the upper portion of the spine.

Torso

The clavicle, ribs, solar plexus, spine, and internal organs are vulnerable to attack.

Groin

This is a very sensitive area.

Extremities

Significant damage to joints and structure can cause immobilization or loss of the use of that limb.

Weapons of the Body

The weapons of the body are divided into two groups: the arms (to include the hands, forearms, and elbows) and the legs (to include the knees, shins, instep, toe, heel, and the ball of the foot). See figure 2-1.

Basic Warrior Stance

The basic warrior stance provides the foundation for all movements and techniques in a close combat situation using the feet, hands, elbows, and chin.

Feet Apart

Place the feet shoulder-width apart, take a half step forward with the left foot, and turn the hips and shoulders at approximately a 45-degree angle. Distribute your weight evenly by bending the knees slightly and adjusting your feet in order to maintain your balance.

Hands Up

Loosen your fists and bring your hands up to chin level or high enough to protect the head without obstructing your vision.

Elbows In

Tuck your elbows in close to your body protecting the torso.

Chin Down

Tuck your chin down taking advantage of the natural protection provided by the shoulders.

Angles of Movement

The purpose of movement is to take control of the confrontation and retain a tactical advantage. Movement is necessary because it—

- Makes the different target areas of the aggressor's body accessible.
- Enables the use of different weapons of your body.
- Increases power and maximize momentum.

When facing an aggressor, movement is made at approximately 45-degree angles to either side of the aggressor. Moving at a 45-degree angle is the best way to avoid an aggressor's strike and put yourself in the best position to attack an aggressor using all of your weapons of the body.

Since MCMAP techniques are initiated from the basic warrior stance shown in figure 2-2, you must know how to move in all directions while maintaining your stance. During any movement, the legs and feet should never cross. Once a movement is completed, you should return to the basic warrior stance. This will help to protect yourself and to put you in the proper position for launching an attack against an aggressor.

Figure 2-2. Basic Warrior Stance.

Marine Corps Martial Arts Program

All movement is initiated by footwork as shown in figure 2-3. Move the first foot that is closest to the direction of movement. Take a 12- to 15-inch step with the first foot, moving your head and body simultaneously. The second foot will rapidly follow the first foot and return to the basic warrior stance. Always end your movement oriented on the aggressor.

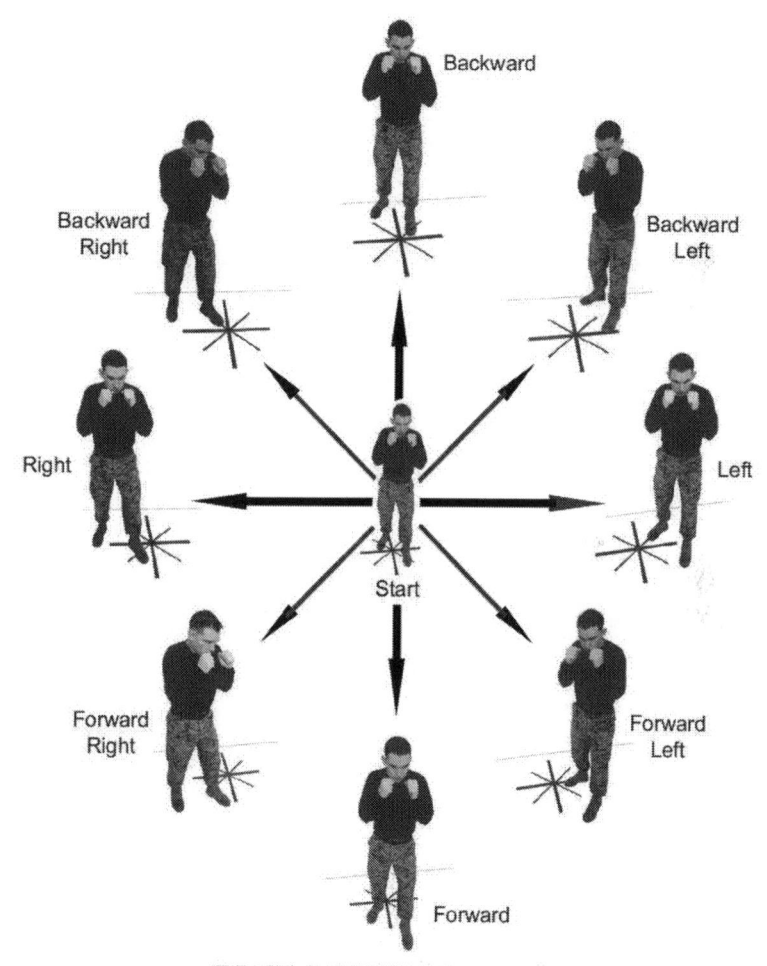

Figure 2-3. Angles of Movement Diagram.

Section II
Punches

Punches are used to stun the aggressor or to set him up for follow-on techniques. It is important to note that while we do not endorse punching in a combative engagement, it is understood that punching is a reflexive behavior. It is important to train in a way that maximizes the damage inflicted upon the aggressor and minimizes the damage to you.

Refer to appendix A for corresponding safeties 1, 2, 3, and 4.

Muscle Relaxation

During instruction on punches, muscle relaxation must be emphasized at all times. The natural tendency during an engagement is to tense up, which results in rapid fatigue and decreased power generation. The person who can remain relaxed during an engagement will generate greater speed, which results in greater power. Relaxing your forearm generates speed and improves reaction time. Clench your fist at the point of impact in order to cause damage to the aggressor and avoid injury to your wrist and hand.

Making a Fist

Punches are executed using the basic fist. When making the basic fist, curl the fingers naturally into the palm of the hand and place the thumb across the index and middle fingers (see fig. 2-4). Do not clench the fist until movement has begun. This increases muscular tension in the forearm and decreases speed and reaction time.

Figure 2-4. Making a Fist.

Just before impact, apply muscular tension to the hand and forearm to reduce injury to you and maximize damage to the aggressor. Contact with the fist should be made with the first two knuckles of the index and middle finger.

When striking with the basic fist, it is important that the first two knuckles are in line with the wrist to avoid injury to the wrist (see fig. 2-5).

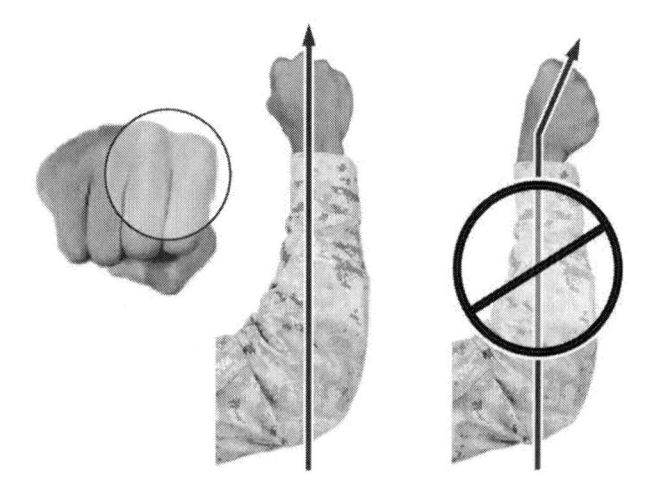

Figure 2-5. Striking Surface.

Weight Transfer

Weight transfer is used to generate power in a punch. This is accomplished by:

- Rotating the hips and shoulders into the attack.
- Moving your body mass forward or backwards in a straight line.
- Dropping your body weight into an aggressor. Your body's mass can be transferred into an attack from high to low or from low to high.

Rapid Retraction

When delivering a punch, rapid retraction of the fist will prevent the aggressor from being able to grab your hand or arm. Once your hand has made contact with the primary target area, quickly return to the basic warrior stance.

Rapid retraction enables you to protect yourself from your aggressor's counterattack by returning your hand and arm to the basic warrior stance. This technique permits the hand and arm to be chambered or placed in the most ideal position for a follow-on, which prepares you to deliver a subsequent punch.

Telegraphing

Telegraphing a strike informs your aggressor that your intentions are to launch an attack through your body movements. Often, an untrained fighter will telegraph his intention to attack by drawing his hand back in view of his aggressor, changing facial expression, tensing neck muscles, or twitching.

These movements, however small, immediately indicate an attack is about to be delivered. If your aggressor is a trained fighter, he may be able to evade or counter your attack. If your aggressor is an untrained fighter, he may be able to minimize the effect of your attack. Staying relaxed helps to reduce telegraphing.

Lead Hand Punch

The lead hand punch is a snapping straight punch executed by the forward or the lead hand. It is a fast, unexpected punch designed to stun an aggressor and to set up for follow-on techniques. A lead hand punch conceals movement and allows you to get close to the aggressor. If possible, lead hand punches should strike soft tissue areas.

Striking Surface

The striking surface is the first two knuckles of the fist.

Target Areas of the Body

Soft tissue areas such as the nose, the jaw, and the throat are the primary target areas.

Technique

~ Assume the basic warrior stance.

~ Snap your lead hand out to nearly a full extension, while rotating your palm to the deck. Do not over extend your elbow because this can cause an injury due to hyperextension of the joint.

~ Keep your rear hand in place to protect your head.

~ Retract your hand rapidly, returning to the basic warrior stance.

~ The fist will travel out and back in a straight line. A mistake that is commonly made is pulling back low and then resetting, which will leave you open to a counter punch.

See figure 2-6.

Figure 2-6. Lead Hand Punch.

Rear Hand Punch

The rear hand punch is a snapping punch that is executed by the rear hand. This is a power punch that is designed to inflict the maximum damage on your aggressor. The power comes from pushing off of your rear leg while rotating your hips and shoulders.

Striking Surface

The striking surface is the first two knuckles of the fist.

Target Areas of the Body

Soft tissue areas such as the nose, the jaw, and the throat are the primary target areas.

Technique
- Assume the basic warrior stance.
- Rotate your hips and shoulders forcefully toward the aggressor and thrust your rear hand straight out, palm down, to nearly full extension.
- Shift your body weight to your lead foot while pushing off on the ball of your rear foot.
- Your rear heel may raise or flare off of the deck.
- Keep your lead hand in place to protect your head.
- Make contact on the aggressor with the first two knuckles of your fist.

~ Rapidly return to the basic warrior stance.
~ The fist travels out and back in a straight line. A common mistake is to pull back low and then reset. This is incorrect and leaves you open to a counterattack.

See figure 2-7.

Figure 2-7. Rear Hand Punch.

Uppercut

The uppercut is a powerful punch originating below the aggressor's line of vision. It is executed in an upward motion traveling up the centerline of the aggressor's body. It is delivered in close and usually follows a preparatory strike that leaves the primary target area unprotected. When delivered to the chin or jaw, the uppercut can render an aggressor unconscious, cause extensive damage to the neck, or sever the tongue.

Striking Surface

The striking surface is the first two knuckles of the fist.

Target Areas of the Body

Soft tissue areas such as the nose, the jaw, and the throat are the primary target areas of the body.

▬ Technique

~ Assume the basic warrior stance.
~ Rotate your fist so that your palm is facing you. Ensure that your lead hand stays up and in place to protect your head.
~ Power is generated from low to high. Start with your body weight low, legs slightly bent.

Marine Corps Martial Arts Program

~ Explode upwards with your legs, hips, and shoulders, drive your fist straight up through the primary target area.
~ Rapidly retract your hand and return to the basic warrior stance.
~ Your fist should never drop below chest level or rise above the aggressor's head when executing this punch to the primary target area of the jaw. A common mistake is to drop the hand all the way to your waist during the execution of the punch, which is incorrect. This execution is used in an effort to get all the power from the arm that is executing the punch. Its power derives from the use of the lower body.

See figure 2-8.

Figure 2-8. Uppercut.

Hook

The hook is a powerful punch that is executed close in and is usually preceded by a preparatory strike.

Striking Surface

The striking surface is the first two knuckles of the fist.

Target Areas of the Body

When delivered to the chin or jaw, the hook can render an aggressor unconscious or cause extensive damage to the neck. When delivered to the body it can cause pain, shortness of breath, and even break ribs.

Chapter 2: Tan Belt — For Official Use Only

Technique

- Assume the basic warrior stance.
- Rotate your rear fist, this will parallel your fist and forearm to the deck.
- Power is generated from side to side by driving with your legs while rotating your hips and shoulders. Your body's rotation drives the fist through your primary target area making contact with the first two knuckles of the fist as your lead hand stays up and in place to protect your head.
- Rapidly retract your hand and return to the basic warrior stance.
- A common mistake is to extend the fist all the way out in an effort to get all of the power from the arm that is executing the punch, which is incorrect. The power derives from the use of the legs and rotation of the upper body.

See figure 2-9.

Figure 2-9. Hook.

Actions of the Aggressor

The aggressor should hold the striking pad tight to the body to avoid injury to his own hands or arms. The striking pad should be held in a position so that the person who is punching connects in the center of it.

Lead and Rear Hand Punches

The striking pad is held with one arm through the center straps and the other upholding the top strap protecting the head as shown in figure 2-10. While you execute the punches the aggressor stands directly in front of you holding the striking pad.

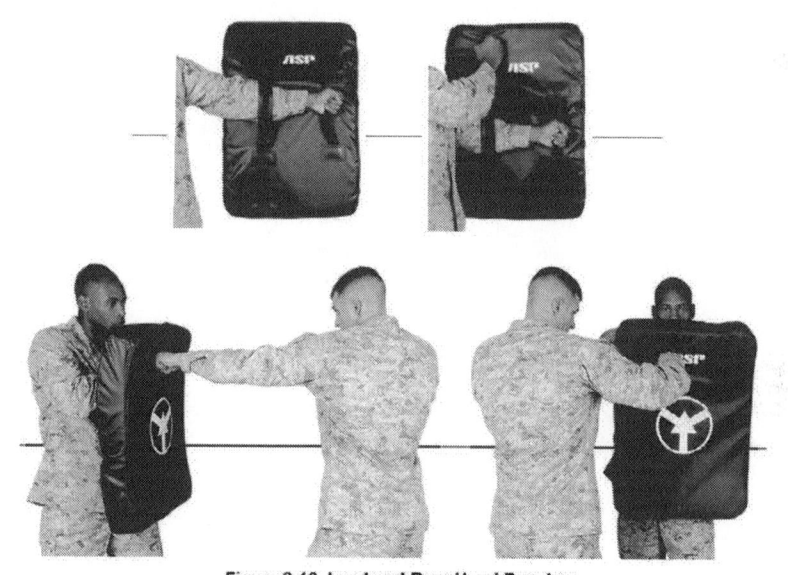

Figure 2-10. Lead and Rear Hand Punches.

Upper Cut

The striking pad is held with one arm through the center straps and the other upholding the top strap. The aggressor will extend his hand holding the top of the striking pad, creating a horizontal striking surface so that you can execute the uppercut (see fig. 2-11).

Hook

The striking pad is held the same as when executing the lead and rear hand punches. The difference is that the aggressor will turn 90-degrees to the side of the incoming hook (see fig. 2-11).

Figure 2-11. Upper Cut and Hook.

Section III
Break-Falls

The purpose of break-falls is to reduce the chance of injury. Break-falls are used to absorb the impact if you should fall or are thrown. This will allow you to quickly get back on your feet and minimize any injury sustained by the fall. This section will cover the front break-fall, back break-fall, left/right side break-fall, and the forward shoulder roll.

Refer to appendix A for corresponding safeties 1, 5, and 6.

Front Break-Fall

A front break-fall is executed to break your fall when falling forward and will never be executed from the standing position during training.

Technique
The front break-fall is taught and practiced in stages, from the deck and from a kneeling position. The front break-fall is never executed from the standing position.

From the Deck
- ~ Begin by lying on the deck on your stomach.
- ~ Place your forearms and palms flat on the deck, with your elbows bent and your chest and head raised off of the deck.
- ~ Raise your head and neck so that you are looking straight ahead.
- ~ Determine the proper hand placement that will allow your forearms and palms to support your upper torso and keep your head off of the deck.
- ~ Proper hand and forearm placement will disperse the impact of the fall and help to keep your head from hitting the deck.
- ~ Practice by raising the forearm and slapping the deck with the fingers extended and joined.

From a Kneeling Position
- ~ Assume a two-knee kneeling position and look up raising your chin.
- ~ Bend your elbows in close to your body and place your palms facing away from you in the position that will allow you to disperse the impact of the fall.

~ Fall forward, breaking your fall with your forearms and palms. The forearms and palms, should strike the deck simultaneously. Fingers will be extended and joined.
~ Keep your head up to avoid striking your chin on the deck.
See figure 2-12.

Figure 2-12. Front Break-Fall.

Back Break-Fall

A back break-fall is executed to break the fall when being thrown or falling backwards and will never be executed from the standing position during training. The back break-fall is taught and practiced in stages, from the deck and from a squatting position.

Technique
The back break-fall is taught in stages, from the deck and from a squatting position.

From the Deck
~ Begin by lying on the deck on your back as shown in figure 2-13.
~ Place your arms at approximately a 45-degree angle downward, out and away from your body, with your palms down.
~ Tuck in your chin.
~ Offer resistance with your arms to raise your head, neck, and shoulders off of the deck.
~ Determine proper arm placement that allows you to keep your head off of the deck.
~ Proper arm placement will disperse the impact of the fall.
~ Once proper arm placement is determined, cross your hands in an X in front of your chest.
~ Tuck in your chin to keep your head up off of the deck.
~ Bend your knees to raise them off of the deck.

Marine Corps Martial Arts Program

~ Swing your arms out and slap the deck, making contact from your forearms to your palms. The forearms and the meaty portion of the palms, down to the fingertips, should strike the deck simultaneously. Your fingers and thumb will be extended and joined.
~ Offer resistance with your arms to raise your head, neck, and shoulders off of the deck.
~ Practice this step as many times as necessary to ensure proper arm placement to break the fall.

From a Squatting Position
~ Squat down on the balls of your feet, tuck your chin in and cross your arms in front of your torso, palms facing you, with fingers and thumb extended and joined.
~ Roll backwards, without pushing off or straightening your legs, keeping your chin tucked to protect your head from the impact.
~ Slap the deck with both extended arms and palms to help disperse the impact. Contact between the deck and your torso and arms should happen simultaneously and your arms should form a 45-degree angle to your body.
~ Keeping your head off of the deck will protect your neck and spinal column from injury.
~ Practice this step as many times as necessary to properly break the fall.

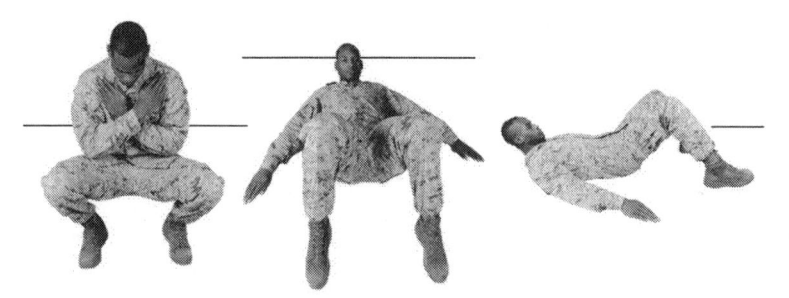

See figure 2-13.

Figure 2-13. Back Break-Fall.

Side Break-Fall

A side break-fall is executed to break your fall, if you should fall on your side.

Technique

It is difficult to execute a break-fall in training because you have to be thrown in order to execute the break-fall properly, however this would create a potentially hazardous condition. The break-fall will be taught and practiced in stages, from the deck, from a squatting position, and from a standing position written for right side, reverse sides for the left.

From the Deck
~ Begin by lying on the deck on your right side. Your right leg should be straight and your left leg should be bent with your foot flat on the deck.
~ Place your right arm, palm facing down, at a 45-degree angle out and away from your body.
~ Raise your head and neck by tucking your chin and tilting your head up away from the deck.

Chapter 2: Tan Belt   For Official Use Only

- ~ Bring your right arm across your body so that your hand is next to your left shoulder with your palm facing inboard.
- ~ With your right arm, slap the deck making contact from your shoulder or forearm down to your hand. Tuck your chin and keep your head raised off of the deck.
- ~ Determine the proper arm placement that allows you to keep your head off of the deck. Proper arm placement will disperse the impact of the fall.
- ~ Practice this step as many times as necessary to determine proper arm placement.

From a Squatting Position
- ~ Assume a one-knee position.
- ~ Bring your right arm across your body so that your hand is next to your left shoulder with your palm facing inboard.
- ~ Fall on your side by sliding your right foot to the left and rolling on your right hip.
- ~ Break your fall with your right arm, slapping the deck, making contact from your shoulder or forearm down to your hand. The arm should strike the deck at a 45-degree angle with respect to your body while simultaneously tucking your chin and keeping your head raised off of the deck.
- ~ To disperse the impact, stretch out your right leg making contact with the deck. Bend your left leg and ensure that your foot makes contact with the deck.
- ~ Practice this step as many times as necessary to ensure proper arm placement to break the fall.

From a Standing Position Written for Right Side, Reverse Sides for the Left
- ~ From the basic warrior stance bring your right arm across your body so that your hand is next to your left shoulder with your palm facing you. Tuck your chin to the left to avoid striking the deck with your head.
- ~ Fall on your right side by sliding your right foot to the left and collapse to the right onto your right thigh, buttocks, and lat muscle.
- ~ Break your fall with your right arm by slapping the deck, making contact from your shoulder or forearm down to your hand. The arm should strike the deck at a 45-degree angle with respect to your body. Keep your head raised off of the deck.
- ~ Everything should make contact with the deck at the same time to disperse the impact. Your right leg is stretched straight to make contact with the deck. Your left leg is bent with the sole of your left foot flat on the deck.

See figure 2-14.

Marine Corps Martial Arts Program

Figure 2-14. Side Break-Fall.

Forward Shoulder Roll

The purpose of the forward shoulder roll is to break a fall from an aggressor's attack and use the momentum to get back on your feet quickly.

Technique

The forward shoulder roll will be practiced in stages, from the kneeling position, and from the standing position.

From the Kneeling Position
- Place your right knee on the deck and your left leg bent with the foot flat on the deck.
- Extend your left arm down through your legs, tucking your chin into your chest. Lower your head and shoulders to the deck.
- Push off with your feet to roll over your left shoulder to your right hip executing the proper finishing position for a right side break-fall.
- With your right leg straight, slap the deck to absorb as much of the impact as possible. Your left leg is bent and the foot hits flat on the deck.

From the Standing Position
- Place your feet shoulder-width apart with your left foot forward.
- Extend your left arm down through your legs, tucking your chin into your chest. Lower your head and shoulders to the deck, but look skyward.
- Push off with your feet to roll from your left shoulder to your right hip and buttock, allowing your forward momentum to bring you back to your feet (**DO NOT DIVE**).
- Rapidly return to the basic warrior stance.

See figure 2-15 on page 2-20.

Figure 2-15. Forward Shoulder Roll.

Section IV
Bayonet Techniques

When executing bayonet techniques, the rifle is held in a modified basic warrior stance. All movement begins and ends with the basic warrior stance.

Grab the pistol grip with your right hand. Keep your trigger finger off of the trigger and include it in the grip. See figure 2-16. While it is possible to execute these bayonet techniques while gripping the buttstock of the weapon, use of the pistol grip allows greater generation of force. These are lethal, offensive techniques that can be used in conjunction with assault fire movement. The small of the buttstock grip is used in nonlethal situations.

With the left hand, grab the handguards of the rifle underhanded. Lock the buttstock of the rifle against the hip with the right forearm. Keep movements of the bayonet blade within a box, shoulder-width across from your neck down to your waistline. The aggressor has a greater chance of blocking your attack if you bring the blade in a wide sweeping movement. Your attacks should close with the aggressor. Bayonet techniques disable or kill an aggressor.

Refer to appendix A for corresponding safeties 1, 2, 4, 7, 8, and 9.

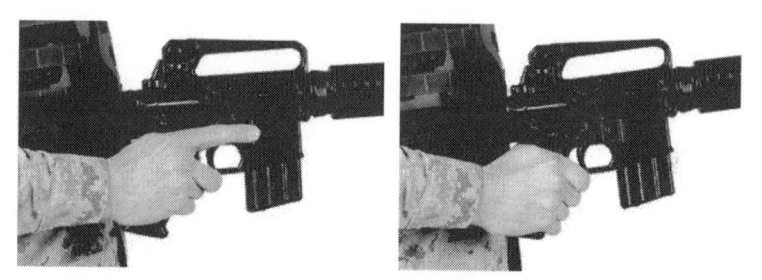

Figure 2-16. Bayonet Techniques.

Straight Thrust

The straight thrust is the most deadly offensive technique because it will cause the most trauma to an aggressor and is the primary offensive bayonet technique. The straight thrust is used to disable or kill an aggressor.

Target Areas of the Body

The primary target areas of the body are the aggressor's throat, groin, or face. The aggressor's chest and stomach are also excellent target areas if he is not protected by body armor or combat equipment.

Technique

- From the modified basic warrior stance step forward with your lead leg, driving off of the ball of your rear foot.
- At the same time, thrust the blade end of the weapon directly toward the aggressor by thrusting both hands forward.
- Retract the weapon and return to the basic warrior stance by stepping forward with the rear foot.

See figure 2-17.

Figure 2-17. Straight Thrust.

Horizontal Buttstroke

The buttstroke is used to weaken an aggressor's defenses, to cause serious injury, or to set him up for a killing blow. It is best executed after a thrust but should always be followed by a slash and a thrust.

Target Areas of the Body

For lethal applications, the head, neck, and unprotected torso are the primary target areas. In a nonlethal situation, the arms, shoulders, and meaty portion of the legs are the primary target areas.

Striking Surface

The strike is executed with the toe of the buttstock of the rifle.

Marine Corps Martial Arts Program 2-23

≡ Technique
~ From the modified basic warrior stance, step forward with your right foot and drive your right elbow forward, parallel to the deck while moving your left hand back toward your left shoulder.
~ Rotate the hips and shoulders into the strike to generate power.
~ Return to the modified basic warrior stance by stepping forward with your left foot and bringing your weapon back executing a slash.

See figure 2-18.

Figure 2-18. Horizontal Buttstroke.

Vertical Buttstroke

The buttstroke is used to weaken the aggressor's defenses, to cause serious injury, or to create space to set him up for a killing blow. It is best executed after a thrust but should always be followed by a slash, thrust, or smash.

Target Areas of the Body

For lethal applications, the head, neck, and unprotected torso are the primary target areas. In nonlethal situations, the arms, shoulders, and meaty portion of the legs are the primary target areas.

Striking Surface

The strike is executed with the toe of the buttstock of the rifle.

Chapter 2: Tan Belt   For Official Use Only

Technique
- From the modified basic warrior stance, step forward with your right foot and drive your right elbow forward, straight up while moving your left hand back toward your left ear.
- Rotate your hips and shoulders, rising slightly, driving with your legs to generate power.
- Return to the modified basic warrior stance by stepping forward with your left foot and bringing your weapon down, executing a slash, or also followed by a smash.

See figure 2-19.

Figure 2-19. Vertical Buttstroke.

Smash

The smash is used as a follow-on technique to the vertical or horizontal buttstroke, primarily when the target was missed on a buttstroke or to gain proper striking distance for close-in engagements.

Target Areas of the Body

The primary target areas of the body are the head, neck, torso, and the arms.

Striking Surface

The strike is executed with the buttstock of the rifle.

Technique
- Start from step one of the vertical buttstroke. Your right foot should be forward with the blade end of the weapon over your left shoulder, weapon roughly parallel to the deck.
- Step forward with your right foot, driving off of your left foot to generate power. Strike the aggressor with the buttstock of the weapon by thrusting the weapon toward your aggressor.

Marine Corps Martial Arts Program 2-25

- ~ As you retract the weapon back to the starting position, take a small step forward with the left leg. **DO NOT JUMP OR HOP**.
- ~ Return to the modified basic warrior stance shown in figure 2-20 by stepping forward with your left foot and executing a slash.

Figure 2-20. Smash.

Slash

The slash is used to cut through the aggressor's defenses or to kill him. It is best to follow up the slash with a thrust to maximize the damage and trauma to the aggressor.

Target Areas of the Body

The primary target areas are the aggressor's head, neck, torso, and the arms.

Striking Surface

The slash is executed with the primary cutting edge of the blade.

Technique
- ~ From the modified basic warrior stance, retract the left hand slightly toward the left shoulder.
- ~ Bring the left hand down and to the right (diagonally) cutting through the target with the blade. To generate more power take a small step with your left foot when you slash, rapidly bringing your right foot back up and return to the modified basic warrior stance.

See figure 2-21 on page 2-26.

Chapter 2: Tan Belt   For Official Use Only

Figure 2-21. Slash.

Disrupt

A disrupt is used as a defensive technique to redirect or deflect an attack in preparation for executing a thrust or other appropriate offensive bayonet techniques. A disrupt is a slight redirection of an aggressor's linear attack such as a straight thrust or a smash.

Technique
- With the weapon locked against the hip with the right forearm, rotate the body to the right or left, moving the bayonet end of the rifle to disrupt the aggressor's attack. Rotation should generate from the hips.
- Contact is made with the bayonet end of the rifle against the barrel or bayonet of the aggressor's weapon.
- Redirect or guide the aggressor's weapon away from your body by exerting pressure against the aggressor's weapon with your weapon. A disrupt should be executed with an economy of motion. You only have to redirect the aggressor's weapon a couple of inches so that the weapon will miss your body.

See figure 2-22.

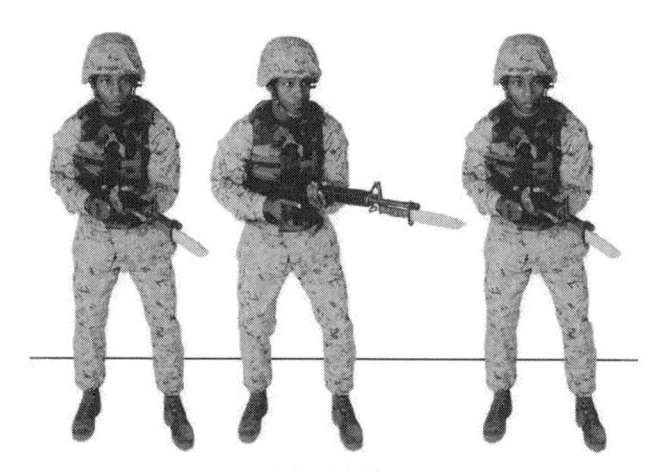

Figure 2-22. A Disrupt.

Section V
Upper Body Strikes

The purpose of upper body strikes is to stun the aggressor or to set him up for a follow up finishing technique. Strikes are unarmed individual striking techniques that are performed with the arms and legs as personal weapons. The hands, forearms, and elbows are individual weapons of the arms and can be used to execute strikes including the hammer fist, knife hand, chin jab, eye gouge, and elbow strikes. These strikes provide a variety of techniques that can be used in any type of close combat encounter.

Refer to appendix A for corresponding safeties 1, 2, 3, and 4.

Principles

Regardless of the strike, there are several principles of execution that ensure its effectiveness.

Generating Power

In executing an effective strike, it is important to generate maximum power through weight transfer:

- Rotate the hips and shoulders into the attack.
- Move your body mass straight forward or backwards in a straight line.
- Drop your body weight into an aggressor. Body mass can be transferred into an attack from high to low or from low to high.

Muscular Tension

There should be muscular tension in the hand and forearm at the moment of impact to maximize damage to the aggressor and to avoid injury to your hand. The arms are relaxed until the moment of impact.

Follow Through

A strike should be delivered so that the weapon such as the hand, or the elbow, hits and remains on the impact target and follows through the target. This technique will inflict maximum damage to the aggressor:

- Strikes with the arms are executed with heavy hands, which means that the strike is executed by driving through with the strike to allow the weight of the hand to go through the primary target area of the body.
- Contact on an aggressor should be made with the arm slightly bent, the arm extends as it moves through the target.
- Using this technique, strikes do not have to be executed at full force to be effective.

Marine Corps Martial Arts Program

Movement

Your movement will put you in the proper position for launching an attack against your aggressor as well as to help protect yourself. Movement is initiated from the basic warrior stance and ends with returning to the basic warrior stance. Each strike can be performed with either the left or right arm depending upon—

- Your angle of attack.
- The position of the aggressor.
- The available vulnerable target areas exposed on the aggressor.

Target Areas of the Body

For each strike, there are target areas on the body: abdominals, ribs, kidneys, chest, and head. A strike to these areas can cause maximum damage to an aggressor. Strikes use gross motor skills as opposed to fine motor skills. The target areas of the body are just that, areas. Pinpoint accuracy on a specific nerve is not needed for the strike to be effective.

Hammer Fist

Striking with the hammer fist concentrates power in a small part of the hand which, when transferred to the target, can have a devastating effect.

Striking Surface

The striking surface of the hammer fist is the meaty portion of the hand below the little finger.

Target Areas of the Body

The primary target areas of the body are the arm and leg joints, the neck, the head, the ribs, and the kidneys.

Angles of Attack

The hammer fist can be thrown horizontally or vertically:

- When thrown horizontally, the hammer fist strike gets its power from hip and shoulder rotation.
- When thrown vertically, the hammer fist strike comes straight down in a straight line and gets its power from dropping your weight into the aggressor by bending the knees and transferring your weight from high to low.

Technique

~ From the basic warrior stance, make a fist. Retract your right hand so that your fist is next to your face and neck. Your arm is bent at approximately a 45- to 90-degree angle. Simultaneously, rotate your right hip and right shoulder backwards.

Chapter 2: Tan Belt   For Official Use Only

- Thrust your fist forward onto the aggressor while rotating your right hip and shoulder forward. Rotate your wrist so that the hammer fist makes contact on the aggressor. Contact should be made with the meaty portion of your hand below the little finger.
- Follow through the primary target area with your fist.
- Return to the basic warrior stance.

See figure 2-23.

Horizontal

Vertical

Figure 2-23. Hammer Fist.

Eye Gouge

The eye gouge is used to attack the aggressor's eyes, blinding him so follow-up strikes can be executed.

Striking Surface

The striking surface is the tips of the fingers and thumb. The strike can be executed by either lead or rear hand.

Target Area of the Body

The primary target area is the eye.

Technique
- From the basic warrior stance extend your left hand with your fingers slightly spread apart to allow entry into the eye sockets.
- With the palm of your hand either toward the deck or toward the sky, thrust your left hand forward into the aggressor's eyes.
- Thrust your hand forward at the aggressor's nose level so that your fingers can slide naturally into the grooves of his eye sockets.
- When striking toward the nose, there is a better chance the fingers will slide up and into the eye sockets.

See figure 2-24.

Figure 2-24. Eye Gouge.

Vertical Elbow Strike (Low to High)

Elbow strikes in general are close range weapons that inflict a great amount of damage due to leverage and the transfer of your body weight.

Striking Surface

The striking surface is two inches above or below the point of the elbow, depending upon your angle of attack, the aggressor's attack angle, and the position of the aggressor.

Target Area of the Body

The chin is the primary target of the vertical elbow strike (low to high).

Angles of Attack

Elbow strikes can be performed from a variety of angles: vertically (low to high or high to low) and horizontally (forward or to the rear).

Technique
- From the basic warrior stance, bend your right elbow, keeping your fist close to your body.
- Your fist is at shoulder level and your elbow is next to your torso.
- Thrust your elbow vertically upward toward the aggressor, keeping your elbow bent throughout the movement.
- Rotate your right shoulder and hip forward and drive up with your legs to generate power.
- Make contact on the aggressor with your right forearm two inches above the point of the elbow.
- Return to the basic warrior stance.

See figure 2-25.

Figure 2-25. Vertical Elbow Strike (Low to High).

Forward Horizontal Elbow Strike

Elbow strikes in general are close range weapons that inflict a great amount of damage due to leverage and the transfer of your body weight.

Target Areas of the Body

The primary target areas are the temple, the spine, the jaw, and the face. By changing the angle of attack slightly you can target the collarbone and other areas depending on your body positioning.

Striking Surface

The striking surface is two inches above or below the point of the elbow, depending upon your angle of attack, the aggressor's attack angle, and the position of the aggressor.

Technique
- From the basic warrior stance bring your elbow up and tuck your right fist near your chest with the palm facing the deck.
- Thrust your right elbow horizontally forward toward the aggressor, keeping your forearm parallel to the deck.
- Keep your fist tucked near your chest with the palm heel facing the deck and your elbow bent throughout the movement.
- Rotate your right shoulder and hip forward.
- Return to the basic warrior stance.

See figure 2-26.

Figure 2-26. Forward Horizontal Elbow Strike.

Actions of the Aggressor

The striking pad should always be held in a position so that the person who is performing the technique connects in the center of the striking pad.

Vertical Hammer Fist

The striking pad is held out in front with both arms out like you are carrying a lunch tray as shown in figure 2-27. Transfer of the hips/torso into all techniques will generate power.

Horizontal Hammer Fist

The striking pad is held close to the body and the aggressor turns 45- to 90-degree angle to accommodate the angle of the strike.

Vertical   Horizontal

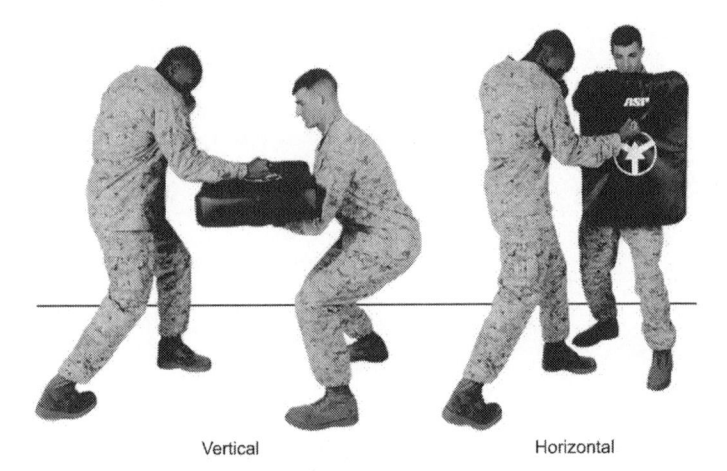

Figure 2-27. Vertical and Horizontal Hammer Fist.

Marine Corps Martial Arts Program

Eye Gouge

The striking pad is held with both hands holding the straps. The aggressor extends his top hand allowing you to execute the technique on the top corner of the striking pad. You should never execute this straight into the striking pad. Take advantage of the slope of the striking pad to minimize the chance of jamming a finger while practicing this technique. See figure 2-28.

Vertical Elbow Strike (Low to High)

The striking pad is held out like the eye gouge to create a horizontal striking surface replicating the aggressor's chin.

Figure 2-28. Eye Gouge.

Horizontal Elbow Strike

The striking pad is held with one arm through the center straps as shown in figure 2-29, and the other arm upholds the top strap to protect the head. The aggressor holding the striking pad stands directly in front of you, while you execute the elbow strikes. The aggressor can change his angle to you up to 90-degrees in order to give you different angles of attack.

Vertical   Horizontal

Figure 2-29. Vertical and Horizontal Elbow Strike.

Section VI
Lower Body Strikes

The purpose of lower body strikes is to stop an aggressor's attack or create an opening in his defense in order to launch an attack. The vertical knee strike, front kick, round kick, and vertical stomp are all lower body strikes that can be performed with either the lead or rear leg. Lower body strikes with the rear leg have greater power because the hips are rotated into the attack. However, the rear leg is further away from the aggressor allowing him a greater reaction time.

Refer to appendix A for corresponding safeties 1, 2, 3, and 4.

Vertical Knee Strike

Knee strikes are excellent weapons for close range combat and are used to create and maximize damage to your aggressor.

Striking Surface

The striking surface is from the top of the knee to two inches above it (not the knee cap).

Target Areas of the Body

If the aggressor is upright, the groin is often the primary target area. If the aggressor is bent over, ideal target areas are the aggressor's face and sternum.

Technique
- From the basic warrior stance, grab the aggressor's neck or gear with both hands, without interlacing your fingers. If you interlace your fingers, your aggressor can pull back or reach behind your head and break the fingers.
- Pull the aggressor down and at the same time raise your right knee driving it up forcefully into the aggressor. Pulling the aggressor down and thrusting the leg upward with your hips generates power.
- Rapidly return to the basic warrior stance.

See figure 2-30.

Front Kick

The front kick is used to stop the aggressor's forward momentum or to set him up for follow-on techniques when the aggressor is in front of you.

Striking Surface

The striking surfaces are the toe of the boot or the bootlaces, depending on the target area.

Marine Corps Martial Arts Program

Figure 2-30. Vertical Knee Strike.

Target Areas of the Body

The primary target areas are the aggressor's groin, knee, shin, and inside thigh.

Technique
- From the basic warrior stance, raise your right knee waist high and pivot your hips into the attack, thrusting your right foot forward toward the aggressor. You may have to shift your body weight to your left leg to maintain balance.
- Never extend your foot above waist high because it is difficult to maintain power and it is easier for the aggressor to counter by blocking or catching your leg. Keep in mind it is difficult to change the direction of a kick after it is initiated because you have limited movement on one leg.
- Make contact on the aggressor with the toe of your right boot or bootlaces.
- Follow through the primary target area with your foot and leg.
- Rapidly return to the basic warrior stance.

See figure 2-31 on page 2-38.

Round Kick

The purpose of the round kick is to cause maximum damage to the knee or to set the aggressor up for follow-on techniques.

Striking Surface

The striking surface is the bottom half of your shin (slightly above the ankle).

Figure 2-31. Front Kick.

Target Areas of the Body

The primary target areas are the aggressor's peroneal nerve (outer portion of the leg), femoral nerve (entire inside of the leg), knee, calf, and Achilles tendon.

Technique
- Raise your rear leg slightly off of the deck, ensuring that the foot of the lead leg is pointed 45-degrees to the outside of the aggressor. With your knee slightly bent, pivot your hips and shoulders into the attack. Thrust your rear leg forward in an arcing motion toward your aggressor.
- You will have to shift your body weight to your left leg to maintain your balance. Ensure that the foot of the lead leg is pointing 45-degrees to the outside of the aggressor.
- With your lead leg slightly bent, extend your rear leg toward your aggressor in an arcing motion. Thrust your rear hip and shoulder forward to generate additional power.
- Make contact on the aggressor with the shin of the rear leg or the top of the foot and follow through the primary target area.
- Rapidly return to the basic warrior stance.

See figure 2-32.

Vertical Stomp

A vertical stomp allows you to remain upright and balanced, in order to rapidly deliver multiple blows with either foot and to quickly and accurately attack your downed aggressor.

Striking Surface

The striking surface is the flat bottom of your boot or the cutting edge of your heel.

Marine Corps Martial Arts Program

Figure 2-32. Round Kick.

Target Areas of the Body

The primary target areas are the aggressor's head or other exposed extremities.

Technique
- From the basic warrior stance, raise the knee of your right foot above waist level. Your right leg should be bent at approximately a 90-degree angle. Shift your body weight to your left leg to maintain your balance.
- Forcefully drive the flat bottom of your right boot or the cutting edge of your right heel down onto the aggressor. At the same time, bend your left knee slightly to drop your body weight into the strike.
- Rapidly return to the basic warrior stance.

See figure 2-33.

Figure 2-33. Vertical Stomp.

Actions of the Aggressor

The striking pad should always be held in a position so that the Marine who is performing the technique connects in the center of the striking pad.

Vertical Knee Strike

The striking pad is held with one arm through the center straps and the other holding the top strap to protect the head. Hold the striking pad tight to your body to avoid injury to your extremities. Keep your top hand up to your head to protect yourself with the striking pad. See figure 2-34.

Front and Round Kicks

Use your hand to hold the top strap on the striking pad. Hold it along the back of your leg so all impact goes in the direction of your bent knee. The strikes should never go against the knee during practice, even with a striking pad. The angle of attack changes between the front and round kicks. The angle of attack for the front kick is vertically forward. The angle of attack for the round kick is horizontally forward.

Figure 2-34. Vertical Knee Strike.

Section VII
Introduction to Chokes

The purpose of chokes is to render your aggressor unconscious or gain control of a close combat situation through less than lethal force. A choke is performed by either closing off of the airway to the lungs, thereby preventing oxygen from reaching the heart, or by cutting off of the blood flow to the brain. Both types of chokes can result in unconsciousness and eventual death for an aggressor. Chokes are classified in two categories: blood chokes and air chokes.

A blood choke is performed on the carotid artery located on both sides of the neck, which carries oxygen-enriched blood from the heart to the brain. When executed properly, a blood choke takes between 8 to 13 seconds for the aggressor to lose consciousness. The blood choke is the preferred choke because its intended effect can be executed quickly, ending the fight.

An air choke is performed on the windpipe or trachea, cutting off the air to the lungs and heart. When executed properly, an air choke takes between two and three minutes for the aggressor to lose consciousness. The air choke is not recommended because of the length of time it takes to stop the fight.

Refer to appendix A for corresponding safeties 1, 2, 6, 10, 11, and 12.

Rear Choke

The rear choke is a blood choke performed when you are behind the aggressor, the aggressor is on the deck, or when you are taking the aggressor to the deck. When teaching the rear choke, do not begin by having students execute the entire technique. Instead, walk the students through the technique, step by step, beginning from a kneeling position and working up to a standing position.

Technique
From a Kneeling Position
- Begin with the aggressor kneeling on the deck and you standing behind him.
- With your right arm, reach over the aggressor's right shoulder and hook the bend of your arm around his neck. Ensure the aggressor's windpipe is positioned within the bend of your arm, but pressure is not being exerted on his windpipe.
- Your chest should be against your aggressor's back.
- With your left hand, clasp both hands together, palm-on-palm, with your right palm facing the deck.
- Exert pressure with your biceps and forearm on both sides of the aggressor's neck on his carotid arteries. Pressure should be exerted with the forearm along the radius bone and the knuckles of the right hand should be facing straight up.
- Ensure that the aggressor's windpipe is positioned within the bend of your arm, but pressure is not being exerted on his windpipe.

~ While maintaining pressure with your biceps and forearm on both sides of the neck, draw the aggressor closer to you by drawing your right arm in.
~ To increase the effectiveness of the choke, apply forward pressure to the back of the aggressor's head with your head by bending your neck forward.

See figure 2-35.

Figure 2-35. Rear Choke from a Kneeling Position.

From the Standing Position
If the aggressor is shorter than you, the procedures are the same as from a kneeling position. If the aggressor is taller than you or the aggressor is wearing bulky gear or a pack on his back, you must get the aggressor in a position where you can reach around his neck and gain leverage to execute the choke.

~ Begin by standing behind the aggressor.
~ Break the aggressor down by reaching over the aggressor's right shoulder with your right arm as you hook his chin, face, or neck with your hand, wrist, or forearm.

Marine Corps Martial Arts Program

- Step or push on the area behind the aggressor's knee with your foot. This will set the aggressor off balance and cause him to bend at the knees and fall forward.
- As the aggressor is brought down, pull back on the aggressor's chin and slide your right forearm around his neck, hooking the bend of your arm around his neck.
- You are now in position to execute the choke. The steps are the same as from the kneeling position.

See figure 2-36.

Figure 2-36. Rear Choke from a Standing Position.

Figure Four Variation

The figure four is a variation of the rear choke, which allows you to gain more leverage on the rear choke. If you cannot secure the rear choke, you may apply the figure four variation to increase the pressure of the choke on the aggressor.

Technique
- Apply a rear choke. Your body should be against the aggressor's body.
- Bring your left arm over your aggressor's left shoulder and grasp your left biceps with your right hand and place your left hand against the back of the aggressor's head.
- With your left hand on the back of your aggressor's head and your elbows in, push the aggressor's head forward and down.
- Draw your right arm in, maintaining pressure with your biceps and forearm on both sides of the aggressor's neck.

See figure 2-37 on page 2-44.

Figure 2-37. Figure-Four Variation.

Marine Corps Martial Arts Program   2-45

Section VIII
Throws

The purpose of a throw is to bring an aggressor to the deck to gain the tactical advantage in a fight. Throws apply the principles of balance, leverage, timing, and body position to upset an aggressor's balance and to gain control by forcing the aggressor to the deck. When executing a throw, it is important to maintain control of your own balance while preventing the aggressor from countering a throw or escaping after he is forced to the deck.

Throw

The throw consists of three parts: entry, off balancing, and execution. The leg sweep is the most basic type of throw in MCMAP that is introduced at the Tan Belt level.

Entry

The first part of a throw is the entry. You want your entry to be quick and untelegraphed to prevent your aggressor from anticipating your movement and countering your attack. You also want to make sure that your body positioning is correct in relation to your aggressor to allow for proper off balancing and execution of the throw.

Off Balancing

The second part of a throw is off balancing. Off balancing techniques are used to control an aggressor by using the momentum of the aggressor to move or throw him. Off balancing techniques can be used to throw an aggressor to the deck while you remain standing or to put yourself in a position for a strike or a choke. Off balancing also aids in the execution of throws, as your aggressor is unable to fight your attack with full strength while off balanced.

Angles of Off Balancing. There are eight angles or directions in which an aggressor can be off balanced. The angles correspond to your perspective, not the aggressor's. Imagine the angles at your feet labeled with forward, rear, right, left, forward-right, forward-left, rear-right, and rear-left. The following are angles that will off balance an aggressor:

- Forward, rear, right, and left are straight angles.
- Forward-right, forward-left, rear-right, and rear-left are considered quadrants, at a 45-degree angle in either direction to your front or your rear.

Off Balancing Techniques. You can off balance the aggressor by pushing, pulling, or bumping him with your hands, arms, or body. Some off balancing techniques are as follows:

- Grabbing an aggressor with your hands and driving him forcefully to one of the rear quadrants or right or left perform pulling.
- Grabbing the aggressor with your hands and driving him forcefully into one of the front quadrants or right or left perform pushing.
- Bumping is executed in the same manner as pushing, but without using your hands to grab the aggressor. Instead, you use other parts of your body such as your shoulders, hips, and legs.

Principles of Off Balancing. Because off balancing techniques rely on the momentum and power generated by the aggressor, they are particularly effective techniques for men and women who may be outsized by their aggressor or the lack of strength that the aggressor has. Off balancing techniques rely on—

- The momentum of the aggressor. For example, if the aggressor is charging at you, you can pull him to drive him to the deck. If the aggressor is pulling you, you can push him to drive him to the deck.
- The generated power of the aggressor. In combat, you are often tired and may be outnumbered. Depending on the generated energy and momentum of the aggressor, you can employ these techniques with very little effort and still obtain effective results.

Practical Application for Off Balancing. Practical application of off balancing will allow the Marine to safely practice off balancing on an aggressor and being off balanced without completing the throw. Practical application is as follows:

- Begin the practical application with students facing one another. Designate one student as the aggressor and the other to perform off balancing.
- Direct the students to do the following; with your left hand, grasp the aggressor's right hand, with your right hand; grasp the aggressor's left shoulder.
- Practice each of the eight angles of off balancing. Ensure that the students push or pull just enough to see that the aggressor is off balanced, not to drive the aggressor to the deck. When the aggressor takes a step back or forward, he is off balanced and compensating to maintain his balance.

Execution

The final part of a throw is the execution. During this portion the aggressor is taken to the deck. This is the defining moment of the throw. Each step before execution is used to set up and assist this final process (see fig. 2-38 on page 2-48).

Leg Sweep

The leg sweep is the first type of throw taught in MCMAP. This throw is used exclusively from the standing position.

Technique
~ Stand facing aggressor in the basic warrior stance.
~ With your left hand, grasp the aggressor's right wrist.
~ Grab the aggressor's clothing or gear if you cannot grab his wrist.
~ Step forward with your left foot on the outside of the aggressor's right foot. At the same time, with your right hand, grasp your aggressor's upper right torso area either on gear or flesh. Your foot should be at least in line or behind the aggressor's foot.
~ Your foot should be placed outside of the aggressor's foot, far enough to provide room to bring the other leg through to execute the sweep.
~ Begin to off balance the aggressor by pulling his wrist downward close to your body and pushing his shoulder backwards.

- ~ When pulling the aggressor's hand, be sure to bring the hand down and close to the trouser pocket.
- ~ Raise your right knee no higher than waist high and bring your foot behind the aggressor's right leg, and stop. The leg should be bent at the knee. This action takes less movement than straightening the leg prior to the sweep. When your leg is raised you should be balanced and in a position to easily off balance the aggressor.
- ~ Sweep through the aggressor's leg, making contact with your calf on the aggressor's calf. At the same time, continue off balancing by pulling the wrist and driving your aggressor back with your right side shoulder.
- ~ In a combative engagement contact will be made with the cutting edge of the heel on the aggressor's Achilles tendon or the calf.
- ~ Bending at the waist, continue to drive through the aggressor's leg as you force him down to the deck. You have to release your grip on the aggressor's shoulder in order to maintain your balance.
- ~ Rapidly return to the basic warrior stance.

See figure 2-38 on page 2-48.

Actions of the Aggressor

The person being thrown should avoid placing too much weight on the right leg (one being swept). Placing all of your weight on the right leg causes you to have to use an undue amount of force to execute the technique that could cause injury to the aggressor. When practicing this technique, allow the person who is executing it to do so while gradually increasing resistance.

Figure 2-38. Leg Sweep.

Section IX
Counters to Strikes

The purpose of counters to strikes is to counter the aggressor's attack and gain the tactical advantage. This section will cover the counter to the rear hand punch and the counter to the rear leg kick.

Refer to appendix A for corresponding safeties 1, 2, 3, 6, 10, 11, 12, 13, and 15.

Counter to Strikes

Regardless of the strike, the counter to a strike requires you to move, block, and strike.

Move

The first step in countering a strike is to move out of the way of the impact of the strike. Movement should both remove you from the point of your aggressor's strike as well as put you in a position to attack. Movement is always initiated from the basic warrior stance and movement is executed at approximately a 45-degree angle to the front or rear. Following movement, return to the basic warrior stance with the toe of your lead foot pointing toward the aggressor.

Block

Different blocks are executed based on the strike. These will be covered with the individual counters.

Strike

Any of the upper body or lower body strikes or combinations of techniques can be executed as a follow-on attack as part of the counter to an aggressor's strike. The choice of follow-on strike depends on your angle to the aggressor, the position of the aggressor, and the available vulnerable target areas that the aggressor has exposed.

Counter to Rear Hand Punch

Counter to rear hand punch is used when the aggressor throws a rear hand punch.

Technique
- Begin with the aggressor in the basic warrior stance extending his right hand in a rear hand punch.
- Step forward and to the left at approximately a 45-degree angle, moving in toward the aggressor. At the end of the movement, return to the basic warrior stance with the left foot forward and the toe pointing toward the aggressor.
- At the same time, raise your left arm and block or deflect the aggressor's rear hand. Do not over extend and reach for the block. Block with the meaty portion of the forearm.

~ Leave your left arm against the aggressor's right arm while stepping forward and to the right at approximately a 45-degree angle to close with the aggressor. Follow through by applying pressure against the aggressor's arm to redirect the strike and, in the process, throw the aggressor off balance. Continuing to step forward will position you to strike an exposed area on the aggressor.

~ Counter with at least three follow-on strikes or techniques to the aggressor's exposed target areas and return to the basic warrior stance.

See figure 2-39.

Figure 2-39. Counter to Rear Hand Punch.

Counter to the Rear Leg Kick

Counter to the rear leg kick is used when the aggressor executes a front kick with his right leg.

Technique

~ From the basic warrior stance, the aggressor should begin by extending his right leg performing a front kick.

~ Step forward and to the left at approximately a 45-degree angle, moving in toward the aggressor.

- At the end of the movement, the left foot is forward with the toe pointing toward the aggressor. At the same time, raise your left arm and block or deflect the aggressor's leg.
- Block with the meaty portion of your forearm. Move out of the way of the strike, which is better than bending down to block the kick.
- Leave your left arm against the aggressor's leg while stepping forward and to the right at approximately a 45-degree angle to close with the aggressor.
- Follow through by applying pressure against the aggressor's leg to redirect the strike, throwing the aggressor off balance. Continuing to step forward will position you so that you can strike an exposed area on the aggressor.
- Counter with at least three follow-on techniques to the aggressor's exposed target areas and return to the basic warrior stance.

See figure 2-40.

Figure 2-40. Counter to the Rear Leg Kick.

Section X

Counters to Chokes and Holds

If executed properly, the counter to the rear choke, the counter to the rear bear hug, and the counter to the rear headlock can render your aggressor unconscious quickly. If a choke is improperly executed it often results in a hold. A hold allows the aggressor control and removes the ability for an attack. It is important that Marines be able to extract themselves from chokes and holds so that they can counterattack and regain the tactical advantage.

There are two principal actions that should be taken to counter any choke—clear the airway and tuck the chin:

- <u>Clear the Airway</u>. A choke can cause unconsciousness in 8 to 13 seconds. Therefore, the first movement in any counter to a choke is to clear your airway so you can breathe. Distracters can be used before or after you have attempted to clear the airway. These techniques include groin strikes, the eye gouge, and foot stomps. Softening techniques are used to loosen an aggressor's hold.
- <u>Tuck the Chin</u>. Once your airway is clear, tuck your chin to prevent the aggressor from re-applying the choke.

Refer to appendix A for corresponding safeties 1, 2, 3, 6, 10, 11, 12, 13, and 15.

Counter to the Rear Choke

The counter to a rear choke is used when the aggressor approaches from the rear and puts his right arm around your throat.

Technique
~ With both hands grab the aggressor's wrist and his forearm at the radial nerve, pull down just enough to clear your airway. Once the airway is clear, tuck your chin to protect your airway and to prevent the aggressor from re-applying the choke. At the same time, drop your body weight down, stepping out with your right leg. This places you more to the aggressor's right side and also makes space for your left foot to step through.

~ With your left foot, step behind the aggressor's right leg keeping both of your legs bent making contact on your aggressor with your left hip placing yourself in almost a squatting position.

~ It is important to keep your legs bent because this places your hips lower than your aggressor's hips so that you can easily off balance him. Bend your legs to the point that you can still maintain your own balance. Your legs need to be in a squatting position with enough balance to maintain control.

~ At the same time, turn forcefully to the left, strike and drive your left elbow into the aggressor's torso while rotating your hips and pivoting to your left. The aggressor should fall to his back or side, causing him to lose his grip.

~ Rapidly return to the basic warrior stance, ready for any follow-on techniques.

See figure 2-41.

Figure 2-41. Counter to the Rear Choke.

Counter to the Rear Headlock

The counter to a rear headlock is used when the aggressor approaches from the rear and puts his right arm around your neck, bending you forward and locking your head against his hip.

▬ Technique
~ Bend forward at the waist. The aggressor faces in the same direction as the student and places his right arm around the student's neck and his forearm across the student's throat.
~ With both hands, grasp the aggressor's wrist and forearm at the radial nerve and pull down to clear your airway. Once the airway is clear, tuck your chin to protect your airway and to prevent the aggressor from re-applying the choke.

~ Take a 12- to 15-inch step forward with your right foot to off balance your aggressor. At the same time reach over your aggressor's right shoulder with your left hand. Once off balancing has been achieved, grab any part of the aggressor's face with the left hand and pull back while rising to a standing position.

~ With your right hand, execute a hammer fist to the aggressor's exposed throat. For safety, the aggressor should bring his free arm up to protect his throat. In addition, contact can be made on your aggressor's chest when executing this technique during practice.

See figure 2-42.

Figure 2-42. Counter to the Rear Headlock.

Counter to the Rear Bear Hug

The counter to the rear bear hug is used when an aggressor approaches from the front or rear. The aggressor will grasp around the arms so that you end up in a vulnerable position with no use of your arms.

Technique

- The aggressor approaches you from behind and applies a bear hug with your arms included in his grasp.
- Drop your body weight down, stepping out with your right leg. This places you more to the aggressor's right side and also makes space for your left foot to step through. Simultaneously hook the aggressor's arms with your hands, slightly flaring your elbows, preventing his arms from slipping off or up into a choke.
- Step behind the aggressor's right leg with your left foot, keeping both legs bent, almost in a squatting position. The left side of your body should be against the aggressor's. Your left hip is in contact with the aggressor's right thigh/buttock, breaking his balance.
- While keeping positive control of your aggressor's arms, turn forcefully to the left, strike and drive your left elbow into the aggressor's torso while rotating your hips and pivoting to your left. The aggressor should fall to his back or side causing him to lose his grip.
- Rapidly return to the basic warrior stance, ready for any follow-on techniques.

See figure 2-43.

Figure 2-43. Counter to the Rear Bear Hug.

Section XI
Unarmed Manipulation

Marines operate within a continuum of force, particularly in support of peacekeeping- or humanitarian-type operations. In these situations, Marines must act responsibly, handling situations without resorting to deadly force. Unarmed restraints and manipulation techniques can be used to control an aggressor. Joint manipulation is used to initiate pain compliance and gain control of an aggressor. Unarmed manipulation consists of the basic wristlock takedown, reverse wristlock, and the armbar takedown. The subject does not actively attack the Marine, but continues to openly defy the Marine's verbal commands. Compliance techniques or distracter techniques can be implemented to remedy behaviors (continued refusal to comply with directions, pulling away, shouting, struggling, locking oneself in a car, or fleeing from the area) that the Marine could encounter at this level. At this level, the physical threat to the Marine remains low.

Refer to appendix A for corresponding safeties 1, 2, 6, 10, and 13.

Compliance Techniques

Compliance techniques are unarmed manipulation techniques used to physically force a subject or aggressor to comply. Compliance can be achieved through close combat techniques of pain compliance using joint manipulation and pressure points. Pain compliance is the initiation of pain in order to gain compliance on the part of the subject.

Distractor Techniques

If you are having difficulty releasing a subject's grip, you can use a softening or distraction technique such as a strike or kick to a pressure point to get the subject to loosen his grip so you can apply a wristlock:

- Add to the effectiveness of the joint manipulation by striking the joint. The strike is executed by driving through with the strike to allow the weight of the hand to go through the primary target area of the body. Strikes do not have to be executed at full force to be effective. The hammer fist strike is an effective softening technique. The hammer fist strike is used to strike the thighs (the femoral and peroneal nerves) and the forearm (the radial nerve).
- Distract the subject so joint manipulation pain compliance can be performed.
- Redirect the movement of the subject or break him down. Striking the forearm in a down and inward movement with a hammer fist strike will cause the subject to bend his elbow so that his direction can be controlled.
- Executing a knee strike or a kick on the inner or outer portion of the subject's thigh can knock a subject off balance or cause him to loosen or weaken his grip. Knee strikes and kicks can be very effective because the subject may never see them coming. A stomp to the foot can also serve as a distraction technique.

Joint Manipulation

Joint manipulation is used to initiate pain compliance and gain control of a subject. Joint manipulation additionally uses the principle of off balancing. A subject can be better controlled when he is knocked off balance. Joint manipulation involves the application of pressure on the joints such as the elbow, wrist, shoulder, knee, ankle, and fingers. Pressure can be applied in two ways:

- In the direction in which the joint will not bend. For example, joints such as the knees and elbows only bend in one direction and when pressure is applied in the opposite direction, pain compliance can be achieved.
- Beyond the point where the joint stops naturally in its range of movement. There are breaking points on each joint. A slow, steady pressure should be applied until pain compliance is reached. Continued pressure will break the joint and may escalate the violence of the situation.

Wristlocks

A wristlock is a joint manipulation that can be applied in a number of ways to achieve pain compliance. The wrist will rotate in a number of directions; it will bend in a single direction until its movement stops naturally. In a wristlock, pressure is exerted beyond the wrist's natural ability to bend or twist the joint. A wristlock can be executed when someone tries to grab you or is successful in grabbing you or your equipment. You can also perform a wristlock when you wish to initiate control of someone.

Basic Wristlock Takedown

Technique
~ Begin with the aggressor placing his hands on your collar/shoulders.
~ Grab the aggressor's left hand with your right hand by placing your thumb on the back of his hand so that your knuckles are facing to your left. Keep the aggressor's left hand centered on and close to your own chest.
~ Hook your fingers across the fleshy part of his palm below the thumb. Rotate the aggressor's palm so that it is now toward him and the fingers point skyward (keeping his palm close to your chest).
~ Bring your left hand up to join the right hand; place the left thumb next to the right on the back of the aggressor's hand. You may place both thumbs on the back of the aggressor's hand, with the thumbs crossed.
~ Hook the fingers of both hands around the fleshy part of the aggressor's palm on both sides of his hand.
~ Apply pressure downward on the back of the aggressor's hand to bend the joint and rotate the wrist outboard, twisting the joint.
~ Apply downward pressure on the wristlock, pivot on the ball of your left foot, and step back with the right foot, quickly turning to your right to take the aggressor to the deck. If you have an aggressor in a wristlock, he can be easily off balanced by pivoting quickly. You can gain better control of the aggressor once he is knocked off balance.

- While you turn in a small circle, the aggressor is forced around in a bigger circle and he cannot move as fast as you and is knocked off balance.
- Continue to apply pressure on the wrist joint as the aggressor lands on his back with his arm straight in the air.
- Slide your lead foot under the aggressor's shoulder.
- Apply pressure with your knee against the aggressor's triceps/elbow while pulling back on his arm and maintain downward pressure on the wrist until pain compliance is achieved.

See figure 2-44.

Reverse Wristlock Takedown

Technique
- Begin with the aggressor placing his right hand on your left collar/shoulder.
- Place the palm of your right hand on the back of the aggressor's right hand and wrap your fingers across the fleshy part of his palm below his little finger.
- Twist the aggressor's hand to the right while placing that hand against your chest. Bring your left hand up, to support your right hand by grabbing the aggressor's hand between both hands, mimic praying. Apply downward pressure on his hand against your chest. Leave the aggressor's hand on the chest to fully control him and to gain leverage. The aggressor's hand should be rotated 90-degrees so that his palm is facing left.
- Step back with your right foot to maintain better balance and lean forward to use body weight to add additional pressure to the joint.

See figure 2-45 on page 2-60.

Armbar Takedown

The armbar takedown is a joint manipulation in which pressure is applied on a locked elbow, at or above the joint, in the direction the joint will not bend. An armbar has to be locked in quickly, but still requires a slow, steady pressure to gain compliance.

Technique
- Face the aggressor and grab his right wrist with your right hand as you step forward to the left, which will be to the right side of the aggressor. Set the aggressor off balance by pulling his wrist to your right hip.
- Pivot on your left foot and step out with your right foot so that you face the same direction as the aggressor. Your right foot must be forward of your left foot.
- Use your left forearm, perpendicular to the aggressor's arm to apply downward pressure on the aggressor's elbow. Maintain control of the aggressor's wrist by keeping it locked into your hip. Apply downward pressure on the arm by dropping your body weight to take the aggressor to the deck. Place your knee onto your aggressor.

See figure 2-46 on page 2-61.

Figure 2-44. Basic Wristlock Takedown.

Figure 2-45. Reverse Wristlock Takedown.

Marine Corps Martial Arts Program

2-61

Figure 2-46. Armbar Takedown.

Chapter 2: Tan Belt

For Official Use Only

Section XII
Armed Manipulations

The purpose of armed manipulation is to ensure that Marines operate within the continuum of force utilizing rifle and shotgun retention techniques. The types of techniques that this section will cover are techniques to counter the muzzle grab, if the aggressor grabs the weapon over or underhand, as well as how to use the weapon correctly to block.

Refer to appendix A for corresponding safeties 1, 2, 8, 9.

Rifle and Shotgun Retention Techniques

With the proper training, Marines will be able to apply techniques so that they can retain their weapons and gain compliance if confronted by individuals who attempts to take away their weapons.

Counter to the Muzzle Grab

This technique is used when you are at port-arms and an aggressor grabs the muzzle of the rifle.

Technique
- Assume the port-arms position.
- Rotate the muzzle in a quick, circular action and then slash downward with the muzzle to release his grip. Rotating the muzzle against the aggressor's thumb is the most effective direction to clear the barrel quickly.
- Step back with your right foot to increase your leverage and balance.

See figure 2-47.

Aggressor Grabs Your Weapon Over Handed

This technique is used when you are at port-arms and an aggressor grabs the handguards of your rifle, palms side down.

Technique
- Assume the port-arm position. The aggressor grabs your handguards with either hand, palm side down.
- Rotate the muzzle in a quick, circular action and then slash downward to release his grip. Rotating the muzzle against the aggressor's thumb is the most effective direction to clear the weapon quickly. Additionally, you may need to step back with your right foot to increase your leverage and balance.

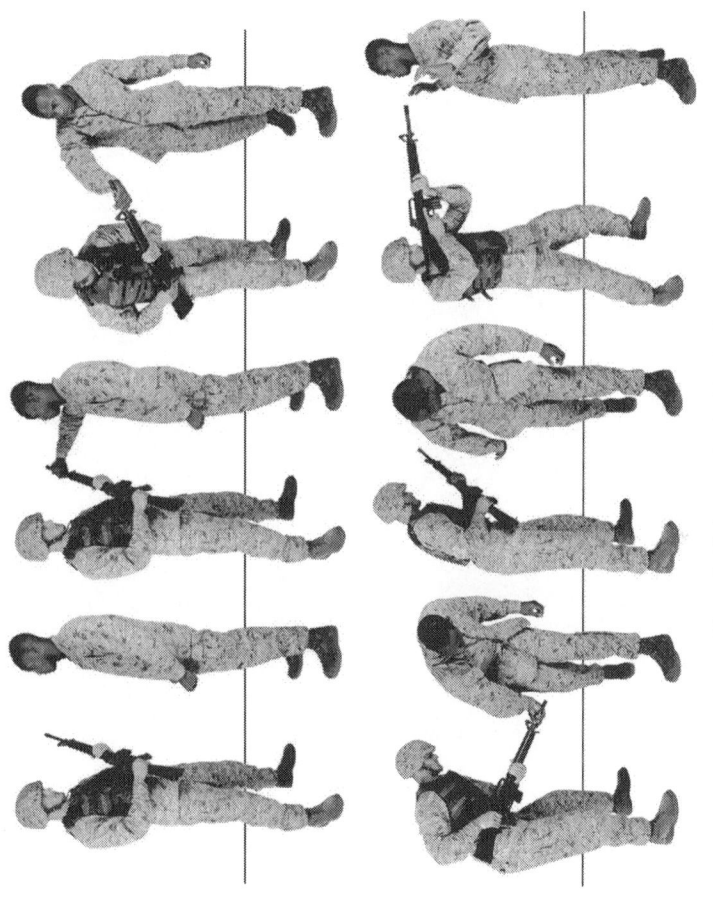

Figure 2-47. Counter to the Muzzle Grab.

~ When the aggressor releases the weapon, step back doubling the distance between you and the aggressor.
~ Execute the ready weapons carry and aim in on the aggressor.

See figure 2-48.

Figure 2-48. Aggressor Grabs Your Weapon Over Handed.

Aggressor Grabs Your Weapon Under Handed

This technique is used when you are at port-arms and an aggressor grabs the handguards of your rifle palms side up.

Marine Corps Martial Arts Program 2-65

Technique

~ Assume the port-arms position. The aggressor grabs your handguards with either hand, palm side up.
~ Rotate the muzzle in a quick, circular action and then slash downward to release the aggressor's grip. Rotating the muzzle against the aggressor's thumb is the most effective direction to clear the weapon quickly. Additionally, you may need to step back with your right foot to increase your leverage and balance.
~ When the aggressor releases the weapon, step back doubling the distance between you.
~ Execute the ready weapons carry and aim in on the aggressor.

See figure 2-49.

Figure 2-49. Aggressor Grabs Your Weapon Underhanded.

Blocks

Blocking techniques are normally executed from the basic warrior stance and are used as a defensive technique to stop an attack. In an engagement, a block can be used if you are out of position and being attacked by an aggressor. In addition, the high block, low block, mid block, and left or right block can be used as primary movements when using the rifle during nonlethal engagements or civil disturbance situations.

The preferred grip for blocks is to hold the weapon at the small of the stock. If the technique is executed while holding onto the pistol grip the blocking surface of the weapon is reduced and there is a greater chance of injury to the hand. If the person lunges at or tries to grab you, block him with your weapon by thrusting it out firmly, with your elbows still bent. Do not try to hit the person with the rifle; the rifle is used as a barrier between you and the person.

High Block

A high block is executed against a vertical attack coming from high to low.

Technique
- Step forward with your lead foot and forcefully thrust your arms up at approximately a 45-degree angle from your body. The weapon should be over the top of your head, parallel to the deck.
- Ensure the weapon is over the head to block a blow to your head, with the pistol grip and magazine facing the attack.
- The elbows are bent but there is enough muscular tension in the arms to absorb the impact and deter the attack.
- Left hand grip will be firm on the handguards. The grip will not change from port-arms.

See figure 2-50.

Figure 2-50. High Block.

Low Block

The low block is executed against a vertical attack coming from low to high.

Technique
- Step forward with your lead foot and forcefully thrust your arms down at approximately a 45-degree angle from your body. The weapon should be at or below your waist, parallel to the deck.
- The elbows are bent with enough muscular tension in the arms to absorb the impact and deter the attack.

See figure 2-51.

Marine Corps Martial Arts Program

Figure 2-51. Low Block

Mid Block

The mid block is executed against a linear/straight attack coming directly toward you.

Technique
~ Step forward with your lead foot and forcefully thrust your arms straight out from your body. The weapon should be held at a position similar to "present arms."
~ The elbows are bent but there is enough muscular tension in the arms to absorb the impact and deter the attack.

See figure 2-52.

Figure 2-52. Mid-Block.

Left or Right Block

A left or right block is executed against a horizontal buttstroke or a slash.

Technique
- Step forward-right or forward-left, at a 45-degree angle, and forcefully thrust your arms to the right or left, holding the rifle vertically in the direction of the attack.
- The elbows are bent with enough muscular tension in the arms to absorb the impact and deter the attack.

See figure 2-53.

Figure 2-53. Left or Right Block.

Section XIII
Knife Fighting

The purpose of knife fighting is to cause massive trauma and damage to an aggressor by executing the vertical thrust or the vertical slash techniques. In any confrontation, the parts of the aggressor's body that are exposed or readily accessible will vary. The goal in a knife fight is to attack soft body vital targets that are readily accessible such as the face, the sides and front of the neck, and the lower abdomen or groin. The extremities function as secondary target areas.

Refer to appendix A for corresponding safeties 1, 2, 3, and 14.

Principles of Knife Fighting

When knife fighting, always execute movements with the knife blade within a box, shoulder-width across from your neck down to your waistline. The aggressor has a greater chance of blocking your attack if you bring the blade in a wide sweeping movement to the aggressor. Your attacks should close with the aggressor, coming straight to your target. Always keep the knife's blade tip forward and pointed toward the aggressor. In each of the knife techniques, apply full body weight and power. In preparation for a vertical slash or a vertical thrust, full body weight should be put into the attack in the direction of the blade's movement. Applying constant forward pressure with your body and blade will keep the aggressor off balance.

Angles of Attack

There are six angles from which an attack with a knife can be launched:

- Vertical strike coming straight down on the aggressor.
- Forward diagonal strike coming in at a 45-degree angle to the aggressor.
- Reverse diagonal strike coming in at a 45-degree angle to the aggressor.
- Forward horizontal strike coming in parallel to the deck.
- Reverse horizontal strike coming in parallel to the deck.
- Forward thrust coming in a straight linear line to the aggressor.

Target Areas of the Body

The goal in a knife fight is to attack soft body, vital target areas that are readily accessible such as the face, the sides and front of the neck, and the lower abdomen or groin. In any confrontation, the parts of the aggressor's body that are exposed or readily accessible will vary. Vital areas are as follows:

- Carotid arteries in the neck are good target areas because they are not covered by body armor or natural protection.
- The lower abdomen and groin region are not covered by body armor.
- The aorta, if not covered by body armor, is an excellent target, which, if struck, can prove fatal in a matter of seconds or minutes.

The extremities are secondary targets. Secondary target areas are those areas that will sever an artery and cause severe bleeding. These target areas are not immediately fatal, but will often become fatal if left unattended. For example—

- The femoral artery, located in the thigh, is a large artery that, when cut, will cause extensive blood loss.
- Attacks on the brachial artery, located between the biceps and triceps on the inside of the arm, can cause extensive bleeding and damage.
- Attacks on the radial and ulnar nerves of the arm can cause extensive bleeding and damage.

Movement

You can move anywhere within a 360-degree circle around an aggressor to gain a tactical advantage and make accessible to you different target areas of your aggressor's body. However, the worst place to be in a confrontation is directly in front of an aggressor. The aggressor can rely on his forward momentum and linear power to create a tactical advantage.

When facing an aggressor, movement is made in a 45-degree angle to either side of the aggressor. Moving at a 45-degree angle is the best way to avoid an aggressor's strike and to put you in the best position to attack an aggressor.

Knife Placement

When a Marine is issued a rifle he probably has been issued a bayonet. If the Marine has been issued a pistol then he probably has been issued a fighting knife. In either case, the knife must be worn where it is easily accessible and where it can best be retained. The specific location of the knife is as follows:

- It is recommended the knife be worn on the weak side hip, blade down. The fighting knife should be placed so its blade is facing forward.
- The knife should be placed behind the magazine pouch where it is easily accessible but not easy for the aggressor to grab.
- Do not place the knife next to something that can cover it like a canteen because the canteen can slide on the cartridge belt, covering the knife and making it inaccessible.

Grip

Your grip on the knife should be natural. Grasp the knife's grip with your fingers wrapped around the grip naturally as it is pulled out of its sheath. This is commonly known as a hammer grip. From this position, the blade end of the knife is always facing the aggressor.

Marine Corps Martial Arts Program

Stance

The basic warrior stance serves as the foundation for initiating knife techniques and is executed as follows:

- The left hand serves as a vertical shield protecting the ribs or the head and neck.
- The right elbow is bent with the blade pointing forward toward the aggressor's head. This position serves as an index point, from which all techniques are initiated.
- The weapon should be held at a level between the top of the belt and the chest.
- The weapon should be kept in close to the body to facilitate weapon retention.

Vertical Slash

Vertical slashing techniques are used to close with the aggressor and can also be a distraction to the aggressor. When executing a vertical slash, the primary target areas of the body are the limbs or any portion of the body that is presented.

Technique
~ Stand facing your aggressor.
~ Thrust your right hand out and bring the weapon straight down on the aggressor.
~ Continue dragging the knife down through the aggressor's body. Maintain contact on the aggressor's body with the blade of the knife. The slashing motion follows a vertical line straight down through the primary target.
~ Return to the modified basic warrior stance.
See figure 2-54.

Figure 2-54. Vertical Slash.

Vertical Thrust

Thrusting techniques are more effective than slashing techniques because of the damage they can cause. However, slashing techniques are used to close with the aggressor in order to get in proximity where a thrusting technique can be used. The thrusting motion follows a vertical line straight up through the primary target. When executing a vertical thrust, the primary target areas of the body are low into the abdomen region or high into the neck.

Technique
- Stand facing your aggressor in the modified basic warrior stance.
- Thrust your right hand toward the primary target, inserting the knife blade straight into the aggressor.
- Pull the knife out of the aggressor.
- Return to the modified basic warrior stance.

See figure 2-55.

Figure 2-55. Vertical Slash.

CHAPTER 3

Gray Belt

Gray Belt is the second belt ranking within MCMAP. Within two years of qualifying as a Tan Belt, all Marines are expected to advance to Gray Belt. Gray Belt includes the completion of basic fundamentals and introduction to intermediate fundamentals of each discipline and is the minimum requirement to attend the Martial Arts Instructor Course. The purpose and principles remain the same as outlined in Tan Belt.

Gray Belt Requirements.

Prerequisites	Recommendation of reporting senior
	Complete Tan Belt sustainment and integration training
	Complete MCI [Marine Corps Institute] 0337, Leading Marines
Training Hours	Minimum of 20 hours of instruction, excluding remedial practice time and testing
Sustainment Hours	Minimum of 5 hours of sustainment, excluding integration training time and practice time for testing

For Official Use Only

Section I
Bayonet

Through proper training, Marines develop the courage and confidence required to effectively use a bayonet for protection and to destroy the aggressor. While closing on an aggressor, a disrupt or a thrust can be used with a bayonet.

Refer to appendix A for corresponding safeties 1, 2, 8, 9, and 14.

Execution

Disruption is the technique used to create an opening for the Marine when closing with the aggressor. To create that opening, the Marine will execute a technique to bring the aggressor's weapon offline. Entry is the movement made to get inside the aggressor's defense in order to execute follow-on techniques. A movement can be a step forward or a small step to an oblique to get within striking distance of the aggressor. Rely on hips/torso for generating power.

Figure 3-1. Modified Basic Warrior Stance.

Movement

Movement provides a stable attack platform, allowing the Marine to employ his bayonet techniques on, out maneuver, and overwhelm his aggressor. Approaching is used when you have located an aggressor and you are within 20 to 25 yards from the aggressor. Assume the modified basic warrior stance (see fig. 3-1). Bend your back so that you are hunched over the weapon and your chin is tucked to protect your neck, minimizing your own target area. Bend your knees to have a lower center of gravity and decrease your profile. Moving at a fast pace, walk using your legs to absorb the impact of your steps, and ensure that your upper body is not bouncing around as you move. The bayonet must stay locked on the aggressor. Hand placement is crucial for a bayonet trainer as shown in figure 3-2.

Figure 3-2. Hand Placement.

Closing

As you reach the critical distance of 5 to 10 feet, you will use a burst of speed to close the final distance between you and the aggressor using your legs to absorb the impact of your steps. Your stride length will not change from when you were approaching the aggressor. To increase your speed, you will vary the turnover rate of your steps (see fig. 3-3), which may cause the aggressor to hesitate during the engagement and give you the psychological and tactical advantage over him (**DO NOT RUN**).

Figure 3-3. Closing.

Bayonet Techniques Disrupt and Thrust While Closing

Disrupt

A disrupt is a technique that intercepts and redirects the aggressor's attack. This redirection opens a path for your blade to enter your aggressor. The disrupt is a parry while closing and thrusting. If you try to parry and then thrust while closing with your aggressor, you will most likely become entangled with the aggressor. To disrupt the aggressor's attack and clear his weapon, your attack must come at a slight angle to his weapon. The resulting collision will redirect the aggressor's weapon and give your blade a clear path to your primary target.

Thrust While Closing

The straight thrust is performed to disable or kill an aggressor. The thrust is the most deadly offensive technique because it will cause the most trauma to an aggressor. Executing the thrust while closing is even more deadly.

Target Areas of the Body

Target areas of the body are the aggressor's throat, groin, or face. The aggressor's chest and stomach are also excellent target areas if they are not protected by body armor or combat equipment.

Technique

- As you close on a static aggressor, disrupt the aggressor's weapon by attacking at a slight angle. Contact is made with the bayonet end of the rifle against the barrel or bayonet of the aggressor's weapon.
- Redirect or guide the aggressor's weapon away from your body by exerting pressure against the aggressor's weapon with your weapon. It is only necessary to redirect the aggressor's weapon a few inches, enough so that the weapon misses your body. This will give your blade a clear path to your primary target.
- Thrust the blade end of the weapon directly toward the target by thrusting both hands forward.
- Retract the weapon and continue moving forward.

See figure 3-4.

> *Note:* Do not allow students to make contact with one another. See appendix A, safety 7.

Figure 3-4. Bayonet Techniques.

Section II
Upper Body Strikes

The upper body strikes stun the aggressor or set him up for follow-on finishing techniques. The hands, the forearms, and the elbows are the unarmed individual weapons of the arms. These weapons provide a variety of techniques that can be used in any type of close combat situation. This section will cover the chin jab/palm heel strike, knife hand strikes, and elbow strikes.

Refer to appendix A for corresponding safeties 1, 2, 3, and 4.

Chin Jab/Palm Heel Strike

The chin jab/palm heel strike can immediately render an aggressor unconscious and cause extensive damage to the neck and spine.

Striking Surface

The striking surface is the palm's heel of the hand.

Target Area of the Body

The primary target area is the bottom of the aggressor's chin.

Technique
- Begin in the basic warrior stance.
- Bend your right wrist back at a 90-degree angle with your palm facing the aggressor and your fingers pointing up.
- Extend your hand into a flat position with your fingers bent and joined at the second knuckle.
- Keep your right arm bent and close to your body.
- Move forward, to close with the aggressor. You may have to grab the aggressor's elbow or the back of his neck with your left hand to maintain control of him.
- Keeping your right arm bent and close to your side, thrust the palm of your hand directly up under the aggressor's chin.
- At the same time, rotate your right hip forward to drive your body weight into the attack to increase the power of the strike.
- Push off on the ball of the right foot to direct your body weight into the attack from low to high.
- The attack should travel up the centerline of the aggressor's chest to his chin. Contact should be made on the aggressor's chin with the heel of your palm. Follow through the primary target area with your hand and body momentum.
- Rapidly return back to the basic warrior stance.

See figure 3-5.

Figure 3-5. Chin Jab/Palm Heel Strike.

Knife Hand Strikes

Knife hand strikes are used when the primary target area is narrow, for example, to allow for strikes on the neck between body armor and a helmet. The knife hand strike can be executed from one of three angles: outside/forward, inside/reverse, and vertical.

Outside/Forward Knife Hand Strike

The outside/forward knife hand strike is used when the primary target area is narrow and to attack soft target areas of the body such as the sides of the neck. Results of the attack can stun the aggressor allowing you to conduct follow-on techniques.

Striking Surface

The striking surface is the meaty portion of the hand between the bottom of the little finger and the wrist.

Target Area of the Body

The primary target area for the knife hand strike is the neck.

Technique
~ Assume the basic warrior stance.
~ Execute a knife hand by extending and joining the fingers of your right hand and placing your thumb next to your forefinger (like saluting).
~ Bring your right hand back over your right shoulder and rotate your right hip and right shoulder backwards.

~ Your arm is bent at approximately a 45- to 90-degree angle. Your elbow should be lower than your shoulder.
~ Thrust your knife hand forward (horizontally) into the aggressor while rotating your right hip and shoulder forward.
~ Rotate your wrist so that your palm is up. Contact should be made on the aggressor with the knife-edge of the hand.
~ Follow through the primary target area with your hand and rapidly return to the basic warrior stance.

See figure 3-6.

Figure 3-6. Outside/Forward Knife Hand.

Inside/Reverse Knife Hand Strike

The inside/reverse knife hand strike is primarily used after performing the outside forward knife strike to attack soft target areas of the body. Results of the attack can stun the aggressor, allowing you to conduct follow-on techniques. The striking surface is narrow, allowing strikes on the neck between the body armor and the helmet.

Striking Surface

The striking surface is the cutting edge of the hand, which is the meaty portion of the hand below the little finger extending to the top of the wrist.

Target Area of the Body

The primary target area for the knife hand strike is the neck.

Technique

~ Assume the basic warrior stance.

~ Execute a knife hand by extending and joining the fingers of your right hand placing your thumb next to your forefinger (like saluting).

~ Bring your right hand over your left shoulder. At the same time, rotate your right shoulder and hip forward.

~ Thrust your knife hand forward (horizontally) onto the aggressor while rotating your right hip and shoulder backwards.

~ Rotate your wrist so that your palm is down. Contact should be made on the aggressor with the knife-edge of the hand.

~ Follow through the primary target area with your hand and rapidly return to the basic warrior stance.

See figure 3-7.

Figure 3-7. Inside/Reverse Knife Hand.

Vertical Knife Hand Strike

The vertical knife hand strike is primarily used to target the back of the neck and attacking soft tissue areas of the body.

Striking Surface

The striking surface is the meaty portion of the hand between the bottom of the little finger and the wrist.

Target Area of the Body

The primary target area of the body is the back of the neck.

Technique
- Assume the basic warrior stance.
- Execute a knife hand by extending and joining the fingers of your right hand and placing your thumb next to your forefinger (like saluting).
- Bring your right hand back over your right shoulder and rotate your right hip and right shoulder backwards.
- Your arm is bent at approximately a 45- to 90-degree angle. Drop your knife hand downward (vertically) onto the aggressor while dropping your body weight and rotating your right hip and shoulder forward.
- Palm should be facing inboard. Follow through the primary target area with your hand.
- Rapidly return back to the basic warrior stance.

See figure 3-8.

Figure 3-8. Vertical Knife Hand.

Elbow Strikes

Elbow strikes such as the rear horizontal elbow strike and the vertical elbow strike (low to high), are used at close range to inflict a great amount of damage due to the leverage and the transfer of your body weight.

The Rear Horizontal Elbow Strike

The rear horizontal elbow strike is primarily effective when being attacked from the rear. The strike is used to stun the aggressor and allow for follow-on techniques.

Striking Surface

The striking surface is 2 inches above or below the point of the elbow, depending on your angle of attack, the aggressor's attack angle, and the position of the aggressor.

Target Area of the Body

The primary target area of the body is the aggressor's body at the level of your elbow.

Technique
- From the basic warrior stance, look over your shoulder to acquire your primary target.
- Thrust your right elbow horizontally backwards toward the aggressor while taking a slight step backwards to generate power. The right forearm is parallel to the deck, with the palm facing the deck.
- Take a step backwards, in order to bring yourself within striking range of the aggressor. Rotate your left hip forward and your right shoulder backwards to generate additional power.
- Follow through with the strike through the primary target area.
- Rapidly return to the basic warrior stance.

See figure 3-9.

Figure 3-9. Rear to Horizontal Elbow Strike.

Vertical Elbow Strike (High to Low)

The vertical elbow strike is primarily effective if the aggressor is bent over to where you can attack the back of the neck. This can stun the aggressor and allow for follow-on techniques.

Striking Surface

The striking surface is 2 inches above the point of the elbow on the triceps.

Target Areas of the Body

The primary target areas of the body are any soft parts of the body that are accessible.

Technique

- Assume the basic warrior stance.
- Bend your right elbow, keeping your fist close at your ear with your elbow at shoulder level.
- Drop your elbow vertically downward toward the aggressor.
- Keep your elbow bent throughout the movement. Rotate your right shoulder and hip forward at the same time dropping your body weight by bending at the knees to generate additional power.
- Follow through the primary target area with the strike.
- Rapidly return to the basic warrior stance.

See figure 3-10.

Figure 3-10. Vertical Elbow Strike (High to Low).

Actions of the Aggressor

Chin Jab/Palm Heel Strike

The striking pad is held with both hands holding the straps as shown in figure 3-11. The aggressor extends his top hand to allow you to execute the technique on the top part of the striking pad.

Marine Corps Martial Arts Program

Outside/Reverse and Inside/Forward Knife Hand

The striking pad is held close to the body and the bagman turns 45- to 90-degree angle in order to accommodate the angle of the strike shown in figure 3-12.

Vertical Knife Hand and Vertical Elbow Strike (High to Low)

The striking pad is held with both arms out in front like you are carrying a lunch tray. Both hands grab the far strap/handle.

Rear Horizontal Elbow Strike

The striking pad is held with one arm through the center straps and the other holding the top strap to protect the head as shown in figure 3-13. While holding the striking pad the aggressor stands directly behind you as you execute the elbow strikes.

Figure 3-11. Chin Jab/Palm Heel Strike.

Figure 3-12. Outside/Reverse and Inside/Forward Knife Hand.

Figure 3-13. Rear Horizontal Elbow Strike.

Chapter 3: Gray Belt   For Official Use Only

Section III
Lower Body Strikes

The purpose of lower body strikes is to stop an aggressor's attack or create an opening in his defense to launch an attack. Lower body strikes include kicks, knee strikes, and stomps. The lower extremities, groin, and targets below your own waistline are the primary target areas. Never kick high as this jeopardizes your balance and leaves you more vulnerable to a counter-attack. Kicks such as the horizontal knee strike, the sidekick, and the axe stomp can be performed with either the lead leg or rear leg. Kicks with the rear leg have greater power because the hips are rotated into the attack. However, the rear leg is further away from the aggressor allowing him a greater reaction time.

Refer to appendix A for corresponding safeties 1, 2, 3, and 4.

Horizontal Knee Strike

Knee strikes are excellent weapons of the body during the grappling stage or during close range fighting.

Striking Surface

The striking surface is the front of the leg, two inches above the knee. If you are unable to separate your hips from your aggressor, the inside of the knee may be utilized as an alternate striking surface.

Target Areas of the Body

If the aggressor is upright, the inside or outside of the thigh are often the primary target areas.

Technique
- From the basic warrior stance, grab the aggressor's neck or gear with both hands, without interlacing your fingers. If you interlace your fingers, your aggressor can pull back or reach behind your head and break the fingers.
- Pull the aggressor down and at the same time raise your right knee driving it up forcefully into the aggressor. Pulling the aggressor down and thrusting the leg horizontally into the aggressor with your hips generating power. Follow through the primary target area with your knee ensuring to strike the aggressor two inches above the knee.
- Return to the basic warrior stance.

See figure 3-14.

Side Kick

The side kick is executed when the aggressor is on either side of you.

Striking Surface

The striking surface is the outside cutting edge of your boot.

Marine Corps Martial Arts Program

Figure 3-14. Horizontal Knee Strike.

Target Area of the Body

The target area of the body is the aggressor's knee.

Technique
- From the basic warrior stance look to the right and raise your right knee waist high. Thrust your right foot to your right side toward the aggressor.
- You will have to shift your body weight to your left leg to maintain your balance.
- Follow through the target area with your foot and leg.
- Rapidly return to the basic warrior stance.

See figure 3-15.

Figure 3-15. Side Kick.

Axe Stomp

The axe stomp is performed when the aggressor is on the deck in a prone position and you are standing.

Striking Surface

The striking surface is the cutting edge of your heel.

Target Area of the Body

The primary target area is the aggressor's head.

Technique
- From the basic warrior stance, raise the right knee above knee level, keeping your right leg slightly bent.
- Shift your body weight to your left leg to maintain balance. The higher the leg is raised, the more power that can be generated; however, ensure you can maintain your balance.
- Forcefully drive the cutting edge of your right heel down onto your aggressor and bend your left knee to drop your body weight into the strike.
- Keep your right knee slightly bent to avoid hyperextension. At the same time, bend your left knee slightly to drop your body weight into the strike.
- Return to the basic warrior stance.

See figure 3-16.

Figure 3-16. Axe Stomp.

Actions of the Aggressor

Horizontal Knee Strike

The striking pad is held with one arm through the center straps and the other holding the top strap as shown in figure 3-17. Hold the striking pad tight to your body to avoid injury to your extremities. Keep the striking pad low and well in the range of your thighs, which are the primary target areas.

Figure 3-17. Horizontal Knee Strike.

Side Kick

Use your hand to hold the top strap on the striking pad. Hold it along the back of your leg so all impacts go in the direction of your knee bending. The strikes should never go against the knee during practice, even with a striking pad (see fig. 3-18).

Figure 3-18. Side Kick.

Section IV
Front Choke

The purpose of a choke is to render your aggressor unconscious or gain control of a close combat situation through less than lethal force. Chokes are performed by either closing off the airway to the lungs, thereby preventing oxygen from reaching the heart, or by cutting off the blood flow to the brain.

When executed properly, a blood choke takes between 8 to 13 seconds for the aggressor to lose consciousness. The air choke is least preferred because it takes longer to render the aggressor unconscious.

Refer to appendix A for corresponding safeties 1, 2, 6, 10, 11, 12.

The front choke is a blood choke performed when you and your aggressor are facing each other. The front choke employs the aggressor's collar to execute the choke.

Technique
- Begin by facing the aggressor.
- With your right hand, grab the back of the aggressor's right collar, ensure that your palm is facing up.
- Keeping the collar tight in your right palm. Reach under your right arm with your left hand and grab the back of the aggressor's left collar, making certain that your palm is facing up, forming an X with your wrists. Attempt to make your thumbs touch.
- Grab the collar with your elbows facing down, curl your wrist inward and pull down toward your chest. Your left radius bone will cut off the aggressor's left carotid artery. Your right radius bone will cut off the aggressor's right carotid artery.
- Make sure that you apply pressure on the carotid arteries and not on the trachea.

See figure 3-19.

Marine Corps Martial Arts Program 3-19

Figure 3-19. Front Choke.

Section V
Hip Throw

The purpose of the hip throw is to bring an aggressor to the deck gaining the tactical advantage in a fight. If an aggressor is moving toward you to attack, a hip throw can be used to take him to the deck while you remain standing. A hip throw is particularly effective if the aggressor is moving forward or pushing you. Execution of the hip throw uses the aggressor's forward momentum.

Refer to appendix A for corresponding safeties 1, 2, 6, 13, and 15.

There is a minimum of 10 fit ins for each throw during sustaining (see fig. 3-20). Completion of steps 1 through 9 would be considered one fit in. To execute a proper fit in, you should stop the technique just before the execution of throwing the aggressor to the deck.

Fit in
- Stand facing your aggressor.
- With your left hand, grasp the aggressor's right wrist.
- Step forward with your right foot on the inside of the aggressor's right foot. The back of your heel should be next to the center or the toe of aggressor's right foot.
- Step back with your left foot, rotating on the ball of your right foot. The back of your heel should be next to the aggressor's toe. Your knees should be bent.
- At the same time, rotate at your waist, and hook your right hand around the back of the aggressor's body anywhere from his waist to his head, depending on your size. If the aggressor is shorter than you, it may be easier to hook your arm around his head. Hand placement should allow you to control the aggressor and pull him in close to you.
- Your backside and hip should be up against the aggressor.
- Rotate your hip up against the aggressor. Your hips must be lower than the aggressor's. Use your right hand to pull the aggressor up on your hip to maximize contact.
- Pull the aggressor's arm across your body and, at the same time, slightly lift the aggressor off of the deck by bending at the waist, straightening your legs, and rotating your body to your left. If the aggressor cannot be easily lifted, your body positioning is not correct.
- Practice these steps as many times as necessary until you determine the proper body positioning.
- Practice the steps again and, this time, continue this action to force the aggressor around your hip and on to the deck.

See figure 3-20.

If an aggressor is moving toward you to attack you, a hip throw can be used to take the aggressor to the deck while you remain standing. A hip throw is particularly effective if the aggressor is moving toward you or pushing you. Execution of the hip throw uses the aggressor's forward momentum.

Figure 3-20. Hip Throw.

Section VI
Counters to Strikes

The purpose of counters to strikes is to counter the aggressor's attack with the counter to the lead hand punch or the counter to the lead leg kick in order to gain a tactical advantage.

Refer to appendix A for corresponding safeties 1, 2, 3, 6, 10, 11, 12, 13, and 15.

Counter to Lead Hand Punch

This counter is used when the aggressor throws a lead hand punch.

Technique
- Begin with the aggressor extending his left hand in a lead hand punch.
- Execute a forward-right angle of movement moving toward the aggressor. Movement is always made to the outside of the aggressor's attacking arm.
- At the end of the movement, the left foot is forward with the toe pointing toward the aggressor.
- At the same time, raise your left arm and block or deflect the aggressor's attacking arm. Block the punch with the meaty portion of your hand, your palm, or the meaty portion of the forearm.
- Leave your left arm against the aggressor's right arm while stepping forward and to the left at approximately a 45-degree angle to close with the aggressor.
- Follow through by applying pressure against the aggressor's arm to redirect the strike and, simultaneously, throw the aggressor off balance.
- Continuing to step forward will position you to strike an exposed area on the aggressor.
- Counter with a strike to the aggressor's exposed target areas.

See figure 3-21.

Figure 3-21. Counter to Lead Hand Punch.

Counter to Lead Leg Kick

The counter to a lead kick is used when the aggressor executes a front kick with his left leg.

≡ Technique
~ Begin with the aggressor extending his lead leg into a front kick.
~ Step forward using a forward-right angle of movement moving in toward the aggressor. Movement is always to the outside of the aggressor's striking leg.
~ At the end of the movement, the left foot is forward with the toe pointing toward the aggressor. At the same time, raise your left arm and block or deflect the aggressor's leg.
~ Block the kick with the meaty portion of your hand, your palm, or the meaty portion of the forearm. Move out of the way of the strike it is better than bending down to block the kick.
~ Continuing to step forward will position you to strike an exposed area on the aggressor.
~ Counter with a strike to the aggressor's exposed target areas.

See figure 3-22.

Figure 3-22. Counter to Lead Leg Kick.

Section VII
Counters to Chokes and Holds

The purpose of counters to chokes and holds is to be able to remove you from the choke and hold in order for you to counterattack and regain the tactical advantage. This section will cover the counter to the front choke, counter to the front headlock, and the counter to the front bear hug.

Refer to appendix A for corresponding safeties 1, 2, 3 6, 10, 11, 12, 13, and 15.

Counter to Front Choke

The counter to the front choke is used when the aggressor approaches from the front and places both hands around your neck to choke you.

Technique
- Begin with the aggressor facing the student and placing both hands around the student's neck.
- The aggressor should place his thumbs on the student's throat and fingers along the side of the student's neck without applying pressure.
- With your left hand, strike and grasp the aggressor's right forearm where the elbow bends and apply downward pressure on the radial nerve with your fingers. The radial nerve is located along the inside of the forearm along the radius bone.
- This action will loosen the aggressor's grip so that you can clear your airway. With your right hand, execute a chin jab/palm heel strike to the aggressor's chin. The chin jab will be delivered between your aggressor's arms.
- At the same time, generate power into the strike by executing a forward movement bringing your left foot to the outside of the aggressor's right foot and rotating your hips into the strike.

See figure 3-23.

Figure 3-23. Counter to Front Choke.

Counter to Front Headlock

The counter to a front headlock is used when the aggressor approaches from the front and puts his right arm around your neck, bending you forward and locking your head against his hip.

Technique
- Begin by having the student bend forward at the waist.
- The aggressor faces the student and places his right arm around the student's neck, his forearm across the student's throat.
- With both hands, grasp the aggressor's wrist and forearm, pull him down to clear your airway. Maintain control of the aggressor's wrist throughout the move. Once the airway is clear, tuck your chin to protect your airway and to prevent the aggressor from re-applying the choke.
- Move your right hand and arm across the aggressor's torso.
- With your left foot, step forward and to the left at a 45-degree angle. Your left foot should be past your aggressor's right foot and out far enough to bring your right foot through and execute a sweep with your right foot against the aggressor's right leg. At the same time, push against the aggressor's chest with your right arm and shoulder to off balance and generate more power for the sweep.

See figure 3-24.

Figure 3-24. Counter to Front Headlock.

Counter to Front Bear Hug

The counter to a front bear hug is executed when the aggressor approaches from the front and puts both of his arms around your body, trapping your arms to your sides.

Technique
- Drop your body weight by bending at your knees and then spread your feet to maintain your balance. At the same time, thrust your shoulders straight up and slightly flare your elbows while raising your hands.
- With your right hand, grasp the upper portion of the aggressor's torso; with your left hand grasp the lower torso on the aggressor's left side.
- With your left foot, step forward and to the left at a 45-degree angle to the outside of the aggressor's right leg, keeping your left leg bent. Ensure that you step deep enough to off balance the aggressor.
- This movement will begin to off balance the aggressor sending him backwards. Continue to push and pull on the aggressor to keep him off balance.
- Drive your right arm and shoulder forward and, at the same time, bring your right leg forward and sweep the aggressor's right leg making calf-on-calf contact, bringing him to the deck.

See figure 3-25.

Figure 3-25. Counter to Front Bear Hug.

Section VIII
Unarmed Manipulation

The purpose of unarmed manipulations is to teach individuals how to operate within the continuum of force, particularly in support of peacekeeping- or humanitarian-type missions. The wristlock come-along, the takedown from a wristlock come-along, takedown from a wristlock come-along and double flexi cuff, escort position, and escort position takedown and single flexi cuff are all forms of unarmed manipulations. In these situations, Marines must act responsibly and handle situations without resorting to deadly force. Unarmed restraints and manipulation techniques can be used to control an aggressor.

Refer to appendix A for corresponding safeties 1, 2, 6, 10, and 13.

Wristlock Come-Along

Technique

- With your right hand, execute a basic wristlock. Incorporate your left hand in a two-handed wristlock in order to gain more control.
- While maintaining pressure on the wrist with your right hand, step forward and to the right and pivot around so that you and the subject are standing next to one another. At the same time, release your right hand and quickly reach under the subject's left arm from behind and grab his hand.
- Your palm and fingers of both hands should be on top of the subject's hand, which is palm side down and your thumbs are across his palm. The subject's elbow should be pointed straight down, with his arm controlled between your forearm and biceps.
- With your left hand, apply downward pressure on the subject's wrist.
- While maintaining downward pressure on the subject's wrist, release your left hand and grab his elbow.
- Apply pressure on the subject's elbow and rotate his elbow up while bringing his wrist down and his hand around to the center of his back.
- Maintain inward and upward pressure on the subject's wrist and elbow throughout to control him.

See figure 3-26.

Figure 3-26. Wristlock Come-Along.

Takedown From a Wristlock Come-Along and Double Flexi Cuff

A takedown is a method used to bring a subject to the deck in order to gain further control of him. This technique is used as a continuation of the wristlock come-along against a subject that is noncompliant.

Technique
- From the basic wristlock come-along, maintain downward pressure on the wrist with your right hand and remove your left hand and place it just above the elbow joint. With your right hand, pull down on the wrist while at the same time use your left hand to pull up on the elbow keeping pressure with the right hand on the wrist, bring the arm into the back.
- With your right foot, push down on the subject's calf or Achilles tendon.
- While maintaining control of the subject's wrist and elbow, apply a slow, steady pressure to bring the subject to the deck.
- Kneel down with both knees on either side of his arm, placing your knees on the subject's back. Apply inward pressure with your knees to lock the subject's arm in place.
- Tell the subject to put his other hand in the middle of his back. Bring the subject's controlled hand to the center of his back.

See figure 3-27.

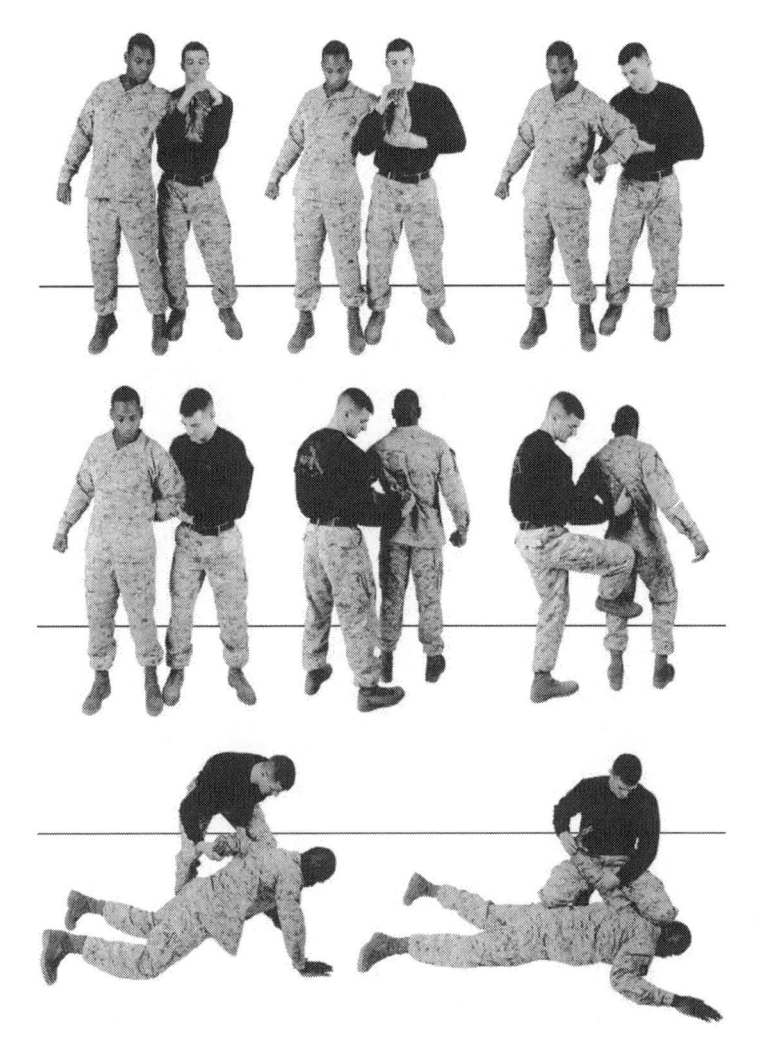

Figure 3-27. Takedown from a Wristlock Come-Along and Double Flexi Cuff.

- Apply the flexi cuffs by grasping the flexi cuffs with your outside hand in the center of the cuffs with the loops pointing up. Place the cuff with the pinky side up on the subject's controlled wrist. Ensure you maintain pressure on the subject's wrist by controlling the subject's hand.
- Using the proper verbal commands, instruct the subject to place his free hand in the center of his back, look away, and cross his ankles. Take your outside hand and slip it through the free cuff.
- Break down the arm naturally and grab the subject's hand as if you are shaking hands with your outside hand. Slip the cuff on after the shake is complete and properly secure the cuff.

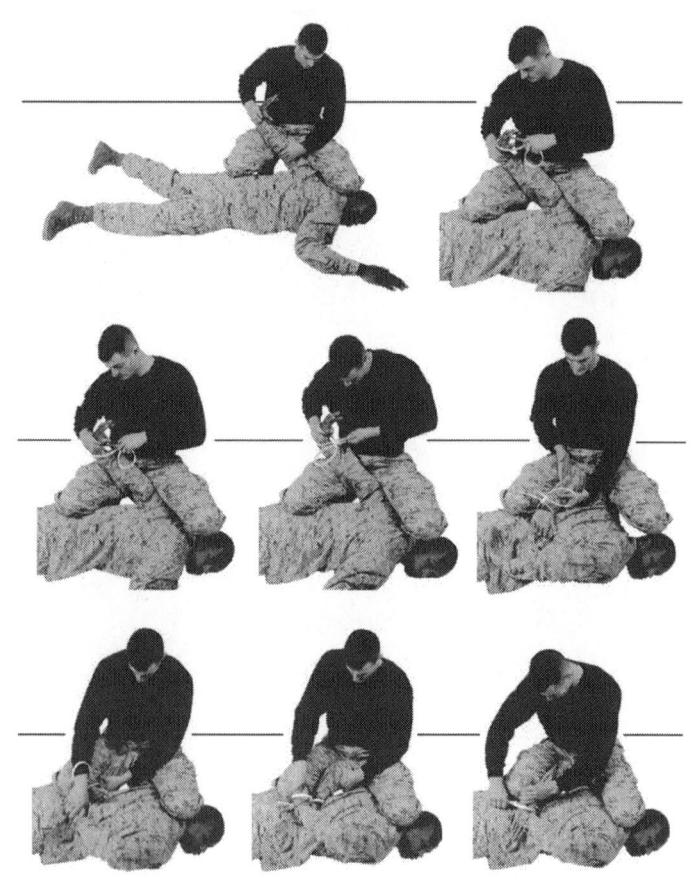

See figure 3-28.

Figure 3-28. Applying Double Flexi Cuffs.

Escort Position

The escort position is used, when necessary, to transport a subject from one location to another.

Technique
- Begin facing the aggressor. Grab the aggressor's right wrist with your right hand and step forward-left. Turn so that you are facing the same direction as the aggressor. Ensure that the aggressor's right palm is facing away from your torso.
- Grab the aggressor's right biceps with your left hand. Your thumb can be used in his armpit on his brachial plexus tie-in for enhanced pain compliance.
- Position the aggressor's right arm diagonally across your torso, keeping his wrist against your right hip. You should be standing to the right of and behind the aggressor. A limited amount of control is achieved by using your chest to place pressure on the aggressor's elbow.

See figure 3-29.

Figure 3-29. Escort Position.

Escort Position Takedown and Single Flexi Cuff

This technique is used as a continuation of the escort position against a noncomplaint subject.

Technique
- From the escort position with the subject to your left, lock the subject's arm straight across your body while rotating his wrist either inboard or outboard away from your body.
- The wrist should be rotated in a direction that will prevent the subject from bending his elbow.
- Rotating your torso away from the subject will allow you to further maintain leverage on the armbar.
- With your left hand or forearm, apply downward pressure above the subject's elbow where the triceps meet, utilizing a straight armbar takedown.

See figure 3-30.

Figure 3-30. Escort Position Takedown and Single Flexi Cuff.

Marine Corps Martial Arts Program 3-37

~ Apply the flexi cuffs by grasping the flexi cuff with your outside hand. Place the cuff on the subject's controlled wrist and grasp the flexi cuff with the ring or pinky finger. Ensure that you maintain pressure on the subject's wrist by controlling the subject's hand.
~ Using the proper verbal commands instructing the subject to place his free hand in the center of his back. Take your outside hand and slip it through the free cuff.

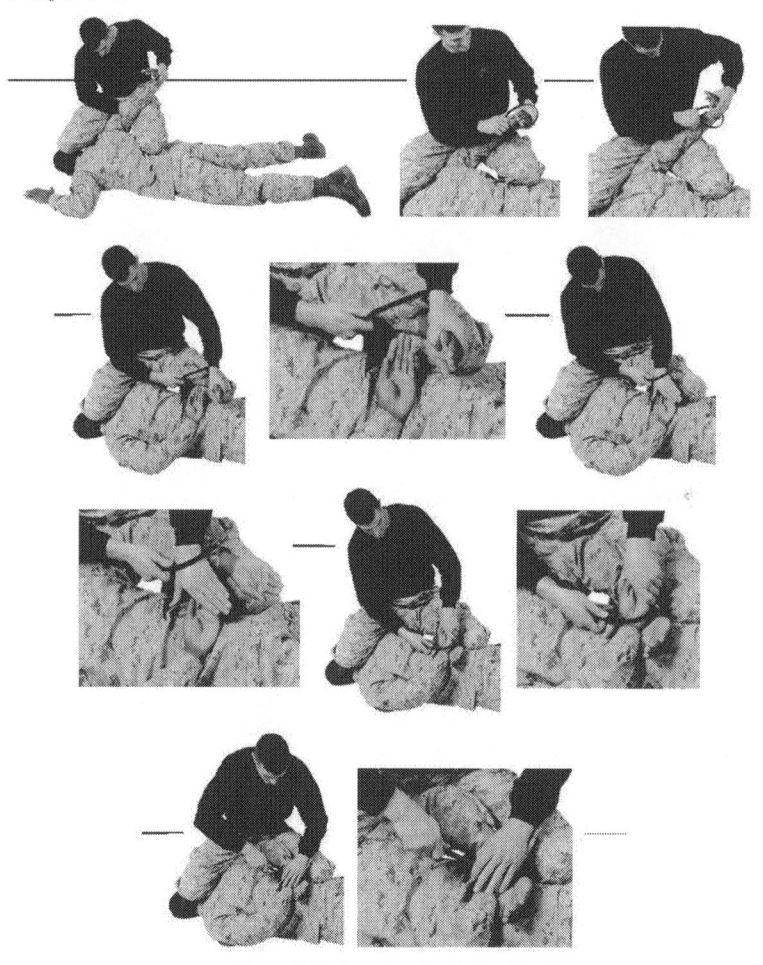

See figure 3-31.

Figure 3-31. Applying Single Flexi Cuffs.

Chapter 3: Gray Belt

Section IX
Armed Manipulation

The purpose of armed manipulation is to ensure that Marines operate within the continuum of force utilizing weapon retention techniques. Weapon retention techniques are used when a Marine is carrying a weapon in a port-arm position and an aggressor attempts to remove the weapon from his grip.

Refer to appendix A for corresponding safeties 1, 2, 8, 9.

Aggressor Grabs With Both Hands (Pushing)

Technique
- Begin with the aggressor grabbing your weapon and pushing on it.
- Utilize the aggressor's momentum and movement by pivoting the body to the left while stepping back with the left foot.
- Throw the person to the deck with a quick jerking movement by lowering the muzzle and swinging the butt of the weapon up, no higher than the shoulders.
- Quickly return to a defensive posture.

See figure 3-32.

Figure 3-32. Aggressor Grabs with Both Hands (Pushing).

Aggressor Grabs With Both Hands Pulling (Stationary)

Technique
- Begin with the aggressor grabbing the weapon and pulling on it.
- Step on the aggressor's foot and push forward to off balance him. This action will drive him to the deck where he can be further controlled.

CAUTION

During training it is essential that you let up on your training partner's foot as soon as his balance breaks. Staying with the foot down on the aggressor's foot as he falls to the deck could cause the ankle or foot to break.

- Quickly return to a defensive posture.

See figure 3-33.

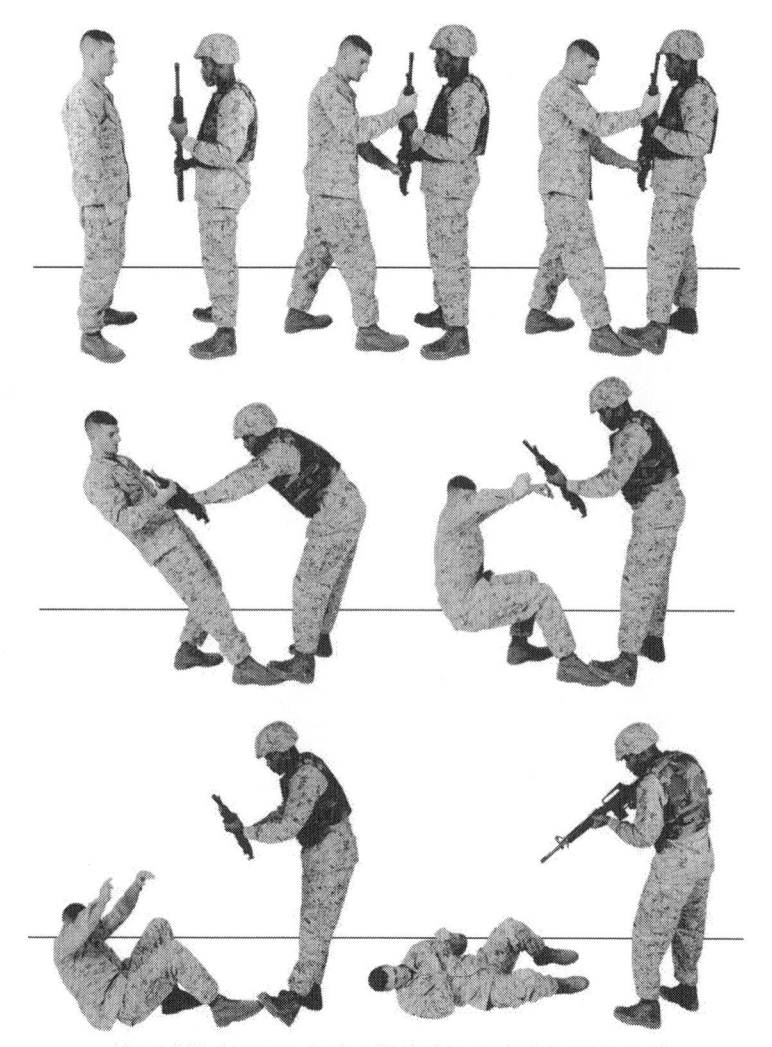

Figure 3-33. Aggressor Grabs with Both Hands Pulling (Stationary).

Aggressor Grabs With Both Hands Pulling (Moving)

Technique
- Begin with the aggressor grabbing the weapon and pulling on it while moving backwards.
- Extend the weapon outward from the body to gain distance.
- Quickly pull the aggressor inward tight to the body and execute a leg sweep on the aggressor's right leg. This action will drive the person to the deck where he can be further controlled.
- When the aggressor releases the weapon, pull the rifle into his shoulder to the ready weapons carry aiming at the aggressor.

See figure 3-34.

Aggressor Grabs Over Handed With Strikes

Technique
- While you are standing at modified port-arms, the aggressor grabs your handguards with the right hand, palm side down. Attempt a counter to the over/under hand grab.
- If unsuccessful, slide your left hand up the handguards; trap the aggressor's finger with your thumb, holding his hand in place with bone pressure on that single digit.
- Attempt to rotate the barrel placing it parallel with the aggressor's forearm.
- If the technique does not work, step forward with the right foot and execute a strike with the toe of the weapon to the aggressor's lead leg. Strikes are to be made to the inside (femoral nerve) or outside (peroneal nerve) of the thigh.
- Complete the counter to the over hand grab technique by stepping back and returning to the basic warrior stance. While maintaining bone pressure on that single digit, rotate the barrel placing it parallel to the aggressor's forearm. Drop your body weight. You can step forward or back in accordance with the aggressor's actions in order to drop body weight.

Marine Corps Martial Arts Program 3-43

Figure 3-34. Aggressor Grabs with Both Hands Pulling (Moving).

Chapter 3: Gray Belt   For Official Use Only

~ Dropping your body weight causes downward pressure to the aggressor's elbow. It will either cause him to release the weapon or act similarly to an armbar.
~ When the aggressor releases the weapon, pull the rifle into your shoulder to the ready weapons carry aiming at the aggressor.

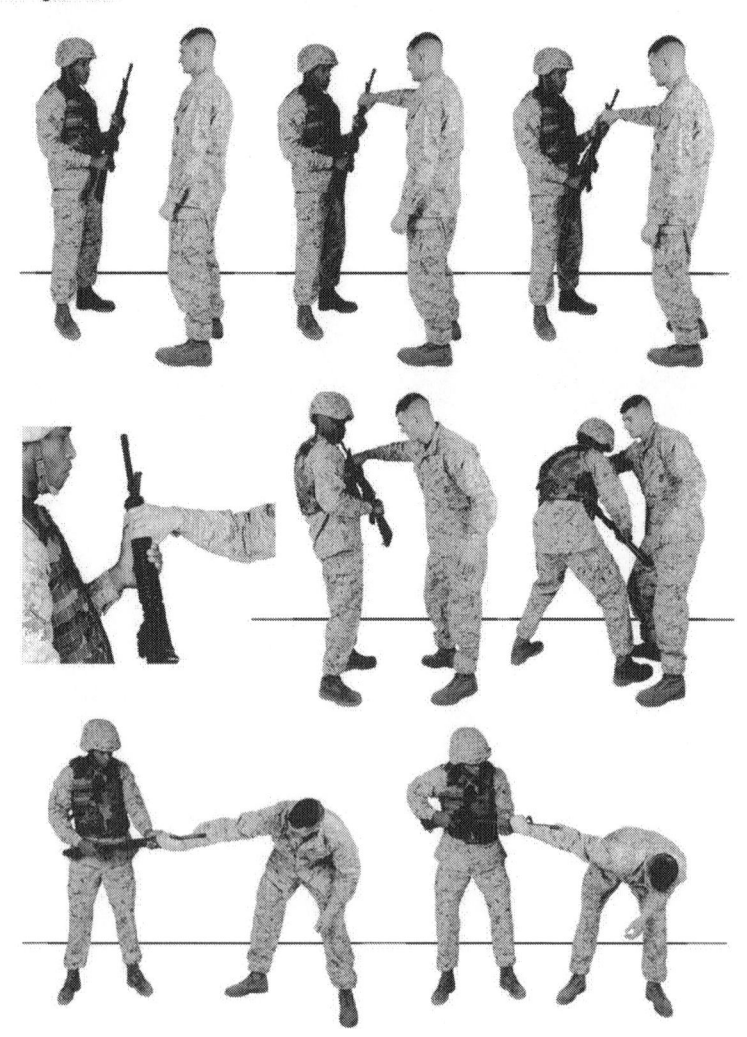

See figure 3-35.

Figure 3-35. Aggressor Grabs Over Handed with Strikes.

Aggressor Grabs Under Handed With Strikes

Technique

- While you are standing at modified port-arms the aggressor grabs your handguards with your left hand, palm side up. Attempt a counter to the over/under hand grab.
- Slide your left hand up the handguards and trap the aggressor's closest finger above the knuckle with your thumb so that he cannot release his grip. Apply bone pressure on the aggressor's finger to initiate pain compliance.
- While maintaining pressure on the aggressor's hand, attempt to execute the counter to underhand grab by lowering the muzzle and raising the buttstock of the weapon no higher than the aggressor's shoulder.
- If the technique does not work, step forward with the right foot and execute a strike with the toe of the weapon to the aggressor's lead leg. Strikes are to be made to the inside (femoral nerve) or outside (peroneal nerve) of the thigh.
- Stepping back with your left foot, pivot on your right, and throw the aggressor past your left side.
- These actions cause the aggressor to release his grip on the weapon or drive him to the deck where he can further be controlled.
- When the aggressor releases the weapon, pull the rifle into his shoulder to the ready weapons carry aiming at the aggressor.

See figure 3-36 on page 3-46.

Figure 3-36. Aggressor Grabs Under Handed with Strikes.

Section X
Knife Techniques

The purpose of knife techniques is to cause enough damage and massive trauma to stop an aggressor by executing the forward thrust, reverse thrust, forward slash, reverse slash, or bulldogging. Thrusting techniques are more effective than slashing techniques because of the damage they can cause. Slashing techniques are used to close with an aggressor. Slashing techniques distract the aggressor or cause enough damage to allow you to close with him and apply more damaging techniques. Targets are usually the limbs or any portion of the body that is presented in proximity where a thrusting technique can be used.

Refer to appendix A for corresponding safeties 1, 2, 3, and 14.

Forward Thrust

The primary objective when fighting with a knife is to insert the blade into an aggressor to cause extensive damage and trauma. A forward thrust follows a linear line straight into the aggressor's neck (high thrust) or abdominal region (low thrust).

Technique
- Stand facing your aggressor. Thrust your right hand toward the target, palm side down, inserting the knife blade straight into the aggressor.
- Once the knife is inserted, twist the blade inward, rotating your palm up. This enables the cutting edge of the blade to be in a position to further cut the aggressor in a follow up action.
- Drop your right elbow and bring the knife to the opposite side of the aggressor's body from where it was inserted. Turn the blade and cut your way out rather than pulling the knife straight out, this causes more damage and trauma to the aggressor. If the aggressor is wearing body armor, it may be difficult or impossible to bring the knife diagonally across his body.

~ At the same time, rotate your hips and shoulders downward to bring your body weight down on the aggressor.
~ Return to the basic warrior stance.

See figure 3-37.

Figure 3-37. Forward Thrust.

Forward Slash

Slashing techniques are used to close with an aggressor. Slashing techniques distract the aggressor or cause enough damage to allow you to close with your aggressor. Targets are usually

Marine Corps Martial Arts Program

the limbs or any portion of the body that is presented. A forward slash is a horizontal, forehand stroke across the target areas of either the neck (high slash) or abdominal region (low slash).

Technique
- Stand facing your aggressor. Extend your right hand to make contact on the aggressor with the knife blade.
- Rotate your palm up to make the blade contact the aggressor. Do not move your arm outside the box, shoulder-width across from your neck to your waistline.
- Continue dragging the knife across the aggressor's body, from your right to your left, in a forehand stroke. Maintain contact on the aggressor's body with the blade of the knife.
- The movement ends with your forearm against your body and the knife at your left hip with its blade oriented toward the aggressor.
- Return to the basic warrior stance.

See figure 3-38.

Figure 3-38. Forward Slash.

Reverse Thrust

The reverse thrust follows a linear line straight into the aggressor's neck a (high thrust) or the abdominal region a (low thrust).

Technique
- Stand facing your aggressor. Bend your right arm, crossing your arm to the left side of your body.
- Thrust your right hand toward the target, palm side up, inserting the knife blade straight into the aggressor.
- Once the knife is inserted, twist the blade by rotating your palm down. This enables the cutting edge of the blade to be in a position to further cut the aggressor in a follow up action.

Chapter 3: Gray Belt

- Bring the knife to the opposite side of the aggressor's body from where it was inserted. Keep the knife inserted; do not remove it.
- Turn the blade and cut your way out rather than pulling the knife straight out, this causes more damage and trauma to the aggressor. If the aggressor is wearing body armor, it may be difficult or impossible to bring the knife diagonally across his body.
- This action can be taken when thrusting to the aggressor's neck or abdomen region. At the same time, rotate your hips and shoulders downward to bring your body weight to bear on the attack.
- Return to the basic warrior stance.

See figure 3-39.

Figure 3-39. Reverse Thrust.

Reverse Slash

Slashing techniques are used to close with an aggressor. Slashing techniques distract the aggressor or cause enough damage to close with your aggressor. Targets are usually the limbs or any portion of the body that is presented. A reverse slash is a horizontal, forehand stroke, across the target areas of either the neck (high slash) or abdominal region (low slash).

Technique
- Stand facing your aggressor.
- Bend your right arm, crossing your arm to the left side of your body. Extend your right hand to make contact on the aggressor with the knife blade.
- Rotate your palm down to place the blade in contact with the aggressor. Do not move your arm outside of the box, shoulder-width across from your neck to your waistline.
- Continue dragging the knife across the aggressor's body, from your left to your right, in a backhand stroke. Maintain contact on the aggressor's body with the blade of the knife.
- Return to the basic warrior stance.

See figure 3-40.

Figure 3-40. Reverse Slash.

Bulldogging

Bulldogging is used to incorporate knife techniques such as thrusting and slashing with movement and open hand strikes.

Technique
- Assume the modified basic warrior stance.
- Maintaining a low silhouette, move forward in a normal walking motion. While moving forward perform a thrusting or a slashing technique.
- After each knife strike, use a strike to the face to stun and distract the aggressor. This is done with an open hand strike with the lead hand.
- Follow up the lead hand strike with one of the thrusting or slashing techniques.
- Have the students practice this combination in sequence while moving forward for a series of repetitions.

See figure 3-41.

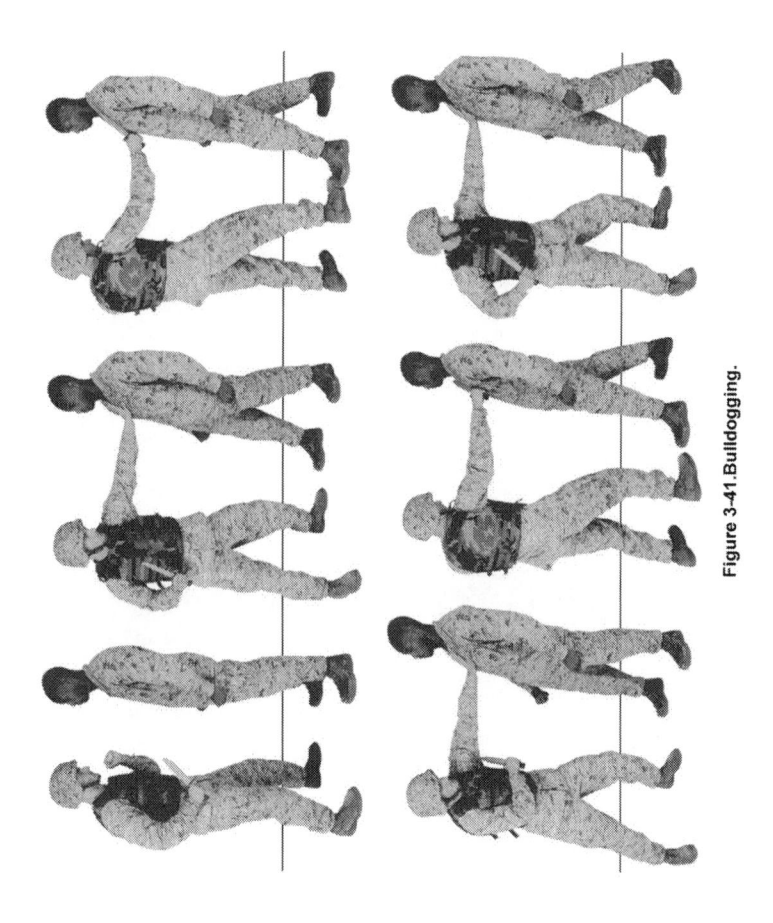

Figure 3-41. Bulldogging.

Section XI
Weapons of Opportunity

In any unarmed close combat situation, a Marine can rely on his body as a weapon by executing the straight thrust, the vertical strike, or a forward strike. In addition, weapons of opportunity enable each individual Marine the creative and innovative ability to utilize any object on the battlefield in order to inflict maximum damage to any aggressor. A Marine should be ready and able to use anything around him to serve as a weapon. This may mean throwing sand or liquid in an aggressor's eyes to temporarily impair his vision so that fatal damage can be done to his head with a rock, e-tool, helmet, or anything that is readily available to the Marine. In a confrontation, a Marine must use whatever it takes to win or face the very real possibility of losing his life.

The principles of employment of weapons of opportunity will depend upon the type of weapon used. Depending upon the shape of the weapon, many of the techniques that are used for employing rifle and bayonet, knife, or baton can be adapted for use with a weapon of opportunity. In cases where this does not apply, the basic principles of angles of attack, stance, movement, and target areas are still applicable.

Refer to appendix A for corresponding safeties 1, 2, 3 4, and 14.

Straight Thrust

The straight thrust technique is effective for executing a frontal attack; it can also be executed as a quick poke to keep a subject away from you, puncture the aggressor's abdominal area, break the aggressor's bones along the rib cage, or cause damage to the aggressor's head or neck.

Technique
- Place your weapon in a two-handed carry. Place your left hand palm side up on the front portion of the weapon, with the rear hand locked into the hip. The weapon should be pulled slightly up with the lead hand.
- Move forward, pushing off with your rear foot to close the distance and generate power. At the same time, thrust your weapon into your aggressor with both hands. The weapon is held either horizontal to the deck or at a slight upward angle, depending on your target area.
- Return to the modified basic warrior stance.

See figure 3-42.

Figure 3-42. Straight Thrust.

Vertical Strike

The vertical strike is effective for executing a frontal attack (high to low).

Technique
~ Begin at the modified basic warrior stance.
~ Rotate your right forearm straight down at the elbow to bring the weapon down on the aggressor. At the same time, forcefully rotate your hips and shoulders toward the aggressor.
~ Shift your body weight to your left foot while pushing off on the ball of your right foot. Drop your body weight into the aggressor from high to low by slightly bending at your back and bending at the knees.

- Follow through with the strike by allowing the weight of the weapon to go through the target area of the body.
- Rapidly returning to the modified basic warrior stance.

See figure 3-43.

Figure 3-43. Vertical Strike.

Forward Strike

The forward strike is effective when executing a frontal attack at a 45-degree angle or horizontally.

=== Technique
- Begin at the basic warrior stance.
- With your left foot, step forward in the direction of the strike. Movement is always made toward the direction of the strike. At the end of the movement, the left foot should be forward with the toe pointing toward the aggressor.
- Rotate your forearm to the right at the elbow to bring the weapon down onto the aggressor. At the same time, forcefully rotate your hips and shoulders toward the aggressor.

- Follow through with the strike by allowing the weight of the weapon to go through the target area of the body.
- Rapidly returning to the modified basic warrior stance.

See figure 3-44.

Figure 3-44. Forward Strike.

Reverse Strike

The reverse strike is effective when executing a frontal attack at a 45-degree angle or horizontally.

= Technique
- Begin at the modified basic warrior stance.
- Execute a forward strike. Your right hand should be near your left shoulder and the weapon extended over your left shoulder. Rotate your forearm to the right at your elbow to bring the weapon down on to the aggressor.

~ At the same time, forcefully rotate your hips and shoulders toward the aggressor. Follow through with the strike by allowing the weight of the weapon to go through the target area of the body.

~ Rapidly returning to the modified basic warrior stance.

See figure 3-45.

Figure 3-45. Reverse Strike.

Section XII
Ground Fighting

The purpose of ground fighting is to provide techniques that allow you to get back to your feet as quickly as possible by executing the counter to the mount position and the counter to the guard position.

Refer to appendix A for corresponding safeties 1, 2, 6, and 10.

Counter to Mount Position

The counter to the mount position is executed if you are lying on your back on the ground and the aggressor is mounted on top of you. The aggressor has the tactical advantage.

Technique
- Begin by lying on your back with the aggressor mounted on top of you.
- Grab the aggressor's gear or clothing on his upper torso and pull him down close to you, while thrusting your hips upward. This will set the aggressor off balance and cause him to extend his arms in front of him and place his hands on the ground to regain his balance.
- With your right arm, over hook your aggressor's arm, from the inside around to the outside, above his elbow.
- Draw your elbow in wrapping it around the aggressors elbow, bending it, bringing him down close to you.

CAUTION
At this point in training, have the aggressor turn his left hand palm side up in order to remove the risk of injury to his wrist.

- Drive the knuckles of your right hand into his chest.
- With your right foot, hook the aggressor's left leg or ankle. This prevents the aggressor from using his leg to prevent you from rolling him over.
- With your left hand, strike the aggressor's side and continue to push him over and roll him off of you to your right side. At the same time, raise your hips and push off with your left leg to assist in rolling the aggressor on his back. Maintain control of the aggressor's hooked arm.
- Avoid being pulled into the aggressor's guard while continuing to maintain control of his over hooked arm, this can be done by bridging at a 45-degree angle over your right shoulder.

~ Go to a squatting position and maintain your aggressor on his side. Keep pressure on the aggressor's hip with your left knee and using your hands, palm-to-palm or figure-four, to apply pressure against the aggressor's left elbow. This action may cause severe shoulder damage. Arching your back as you rise will assist in breaking the aggressor's arm.
~ Return to the basic warrior stance.

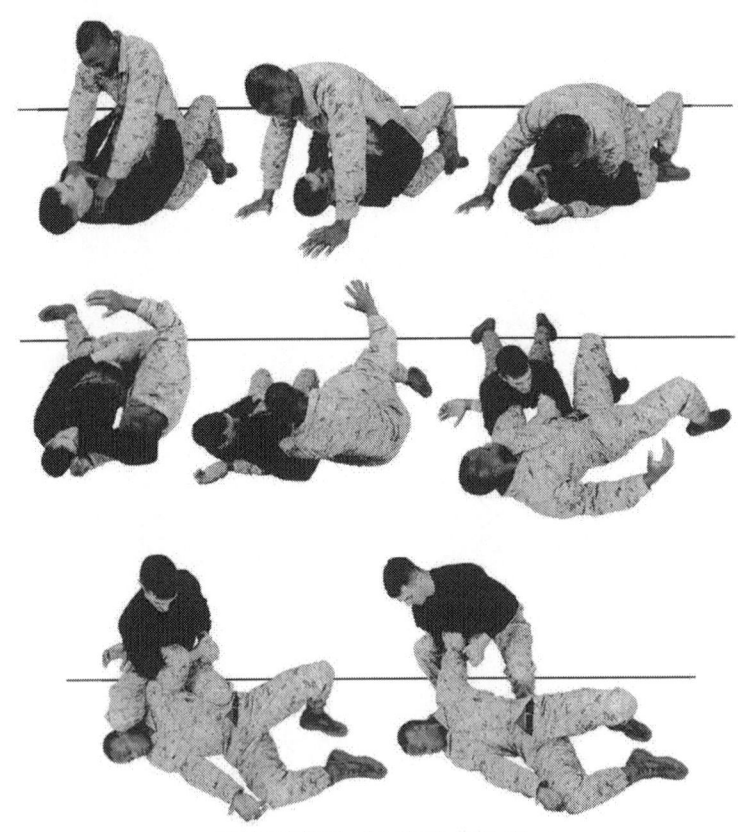

See figure 3-46.

Figure 3-46. Counter to Mount Position.

Counter to Guard Position

The counter to the guard position is executed if the aggressor is lying on his back on the ground and you are kneeling on the ground between his legs with his wrapped around you.

Marine Corps Martial Arts Program	3-61

Technique

- Begin with the aggressor lying on his back on the deck. You are in his guard.
- With your elbows, strike the aggressor's legs at the femoral nerve on the inside of the thigh to drive and separate his legs.
- With your right fist, strike the aggressor's groin.
- At the same time, hook your left arm underneath the aggressor's right knee from the inside, ensure your upper body is staying low and your chin is tucked.
- Quickly throw the aggressor's right leg over your head with your left arm, as you move to your left, and return to the basic warrior stance.

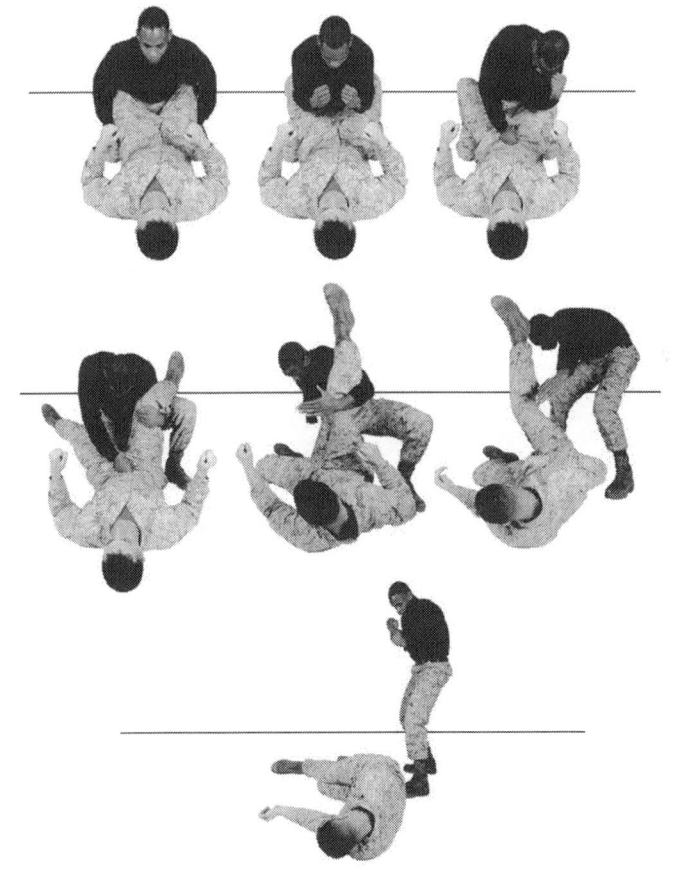

See figure 3-47.

Figure 3-47. Counter to Guard Position.

Chapter 3: Gray Belt	For Official Use Only

This Page Intentionally Left Blank.

CHAPTER 4: Green Belt

Green Belt is the third belt ranking within MCMAP. Within two years of qualifying as a Gray Belt, all Marines are expected to advance to Green Belt. Green Belt includes completion of intermediate fundamentals. Purpose, principles, and movement remain the same as outlined in Gray Belt.

Green Belt Requirements.

Prerequisites	Recommendation of reporting senior
	Complete Gray Belt sustainment and integration training
	Appropriate level PME complete
Training Hours	Minimum of 17.65 hours, excluding remedial practice time and testing
Sustainment Hours	Minimum of 8 hours of sustainment, excluding integration training time and practice time for testing

For Official Use Only

Section I
Bayonet Techniques

When using bayonet techniques you will approach, close, and thrust in order to disable or kill an aggressor. When executing bayonet techniques, all movement begins and ends with the basic warrior stance.

Refer to appendix A for corresponding safeties 1, 2, 4, 7, 8, 9, and 14.

Fundamentals

Mindset

When engaging in combat, mindset will be the determining factor of success or failure, regardless of technical proficiency. Anyone can train in a martial skill, but few have the mind and will to use their skills to kill or injure. Mindset is often the mental trigger in the defining moment that forces you to commit to an aggressor with the goal of injury or death.

Holding the Rifle

When executing bayonet techniques, the rifle is held in a modified basic warrior stance. All movement initiates and ends with the basic modified warrior stance:

- With the right hand, grasp the pistol grip with the trigger finger kept straight and off of the trigger.
- With the left hand, grasp the handguards of the rifle underhanded.
- Lock the buttstock of the rifle against the hip with the right forearm.
- Always execute movements with the bayonet blade at the waistline. The aggressor has a greater chance of blocking your attack if you bring the blade in a wide sweeping movement to the aggressor.
- Your attacks should close with the aggressor, coming straight to your target. Always keep the bayonet end of the rifle oriented toward the aggressor.

Movement

Movement is used to get from one place to another when the threat of contact with the aggressor is imminent. The principles of angles of approach and movement taught in the Tan Belt fundamentals class apply to movement with the rifle and bayonet:

- Assume the modified basic warrior stance.
- Execute movement in the specified direction for one step. Once all of the students have become familiar with all of the movements, they will all execute multiple steps in unison.

Executing Turns

There will be times when it is necessary to turn during an engagement. This is especially true when engaging with multiple attackers. To execute a turn—

- Turn both right and left while keeping the rifle locked into position and the blade oriented to the front. Ensure that you are turning at the torso and not using your arms to move the weapon from left to right.
- Move in a straight line while scanning the area from right to left and left to right by turning your torso and keeping the bayonet oriented in the direction of vision.
- Change the direction of movement to the right by pivoting off of the ball of your left foot as it hits the deck and step with the right foot in the new direction of movement. Ensure that the rifle remains locked into position with the bayonet oriented in the direction of movement.
- Change the direction of movement to the left by pivoting off of the ball of your right foot and step with your left foot in the new direction of movement. Ensure that the rifle remains locked into position with the bayonet oriented in the direction of movement.

Approaching

Approaching is used when you have located an aggressor and you are within 20 to 25 yards from the aggressor. To approach—

- Assume the modified basic warrior stance.
- Bend your knees so that you have a lower center of gravity and so that your profile is decreased.
- Bend at your waist so that you are hunched over the weapon and your chin is tucked to protect your neck, minimizing target area.
- Move at a fast walk using your legs to absorb the impact of your steps. Ensure that your upper body is not bouncing around as you move and the bayonet stays locked on the aggressor.

Closing

Closing is done when you are actively engaged with an aggressor at approximately 5 to 10 feet. Closing is executed to gain the psychological and tactical advantage over the aggressor. To close—

- Assume the modified basic warrior stance.
- Bend your knees so that you have a lower center of gravity and so that your profile is decreased.
- Bend at your waist so that you are hunched over your weapon and your chin is tucked to protect your neck, minimizing target area.
- As you reach the critical distance of 5 to 10 feet, you will use a burst of speed to close the final distance between you and the aggressor using your legs to absorb the impact of your steps.
- Ensure that the upper body is not bouncing around as you move.
- Ensure that the bayonet stays locked on the aggressor. This could cause the aggressor to hesitate during the engagement, which can give you the psychological and tactical advantage.

Execute a Disrupt and a Thrust While Closing With a Moving Aggressor

A disrupt is used as a defensive technique to redirect or deflect an attack in preparation for executing a thrust or other appropriate offensive bayonet techniques. A disrupt is a slight redirection of an aggressor's linear attack followed on by a thrust. The straight thrust is performed to disable or kill an aggressor and is the most deadly offensive technique. The straight thrust will cause the most trauma to an aggressor and is the primary offensive bayonet technique. During training, when performing a disrupt and a thrust while closing with a moving aggressor, have the students execute the movement slowly using another person as the aggressor. When training, have the students execute movement slowly using another Marine as the aggressor.

Target Areas of the Body

The primary target areas of the body are the aggressor's throat, groin, or face. The aggressor's chest and stomach are also excellent target areas if they are not protected by body armor or combat equipment.

Technique

- As you close on a moving aggressor, disrupt the aggressor's weapon by attacking at a slight angle. Contact is made with the bayonet end of the rifle against the barrel or bayonet of the aggressor's weapon.
- Redirect or guide the aggressor's weapon away from your body by exerting pressure against the aggressor's weapon with your weapon. You only need to redirect the aggressor's weapon a few inches, enough so that the weapon misses your body. This will give your blade a clear path to your target.
- Thrust the blade end of the weapon directly toward the target by thrusting both hands forward. Retract the weapon and continue moving forward.

See figure 4-1.

Buttstroke Offline

The buttstroke is used to weaken an aggressor's defenses, to cause serious injury, or to set him up for a killing blow. The strike is executed with the heel of the buttstock of the rifle. It is best executed after a thrust, but if executed as the initial movement, it should be followed by a slash and/or thrust.

Target Areas

For lethal applications, the head, neck, and unprotected torso are the primary target areas of the body. In a nonlethal situation, the arms, shoulders, and the meaty portion of the legs can be target areas.

Marine Corps Martial Arts Program 4-5

Striking Surface

The strike is executed with the toe of the buttstock of the rifle.

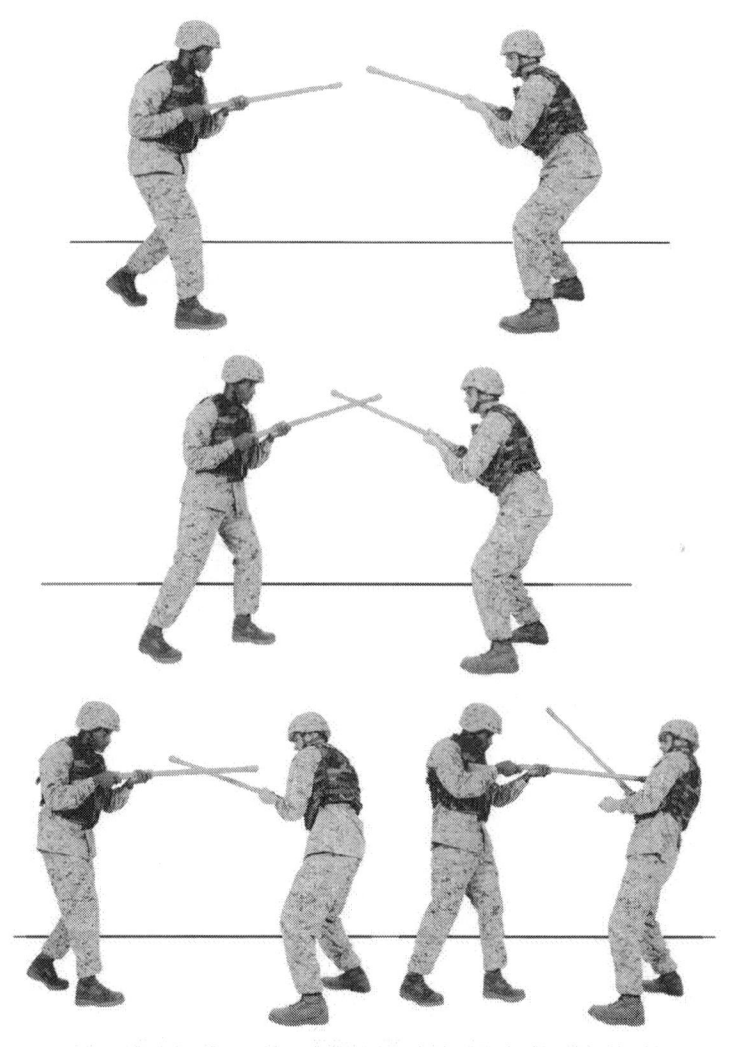

Figure 4-1. Disrupt and Thrust While Closing with a Moving Aggressor.

Chapter 4: Green Belt For Official Use Only

Technique

- Have the student execute movement slowly using another Marine as an aggressor.
- Execute your approach and close on the aggressor.
- While closing, execute a straight thrust. This thrust is interrupted by the aggressor's weapon (e.g., front-sight assemblies catching on each other).
- Once the aggressor's weapon has been interrupted, move forward-right while executing a horizontal or vertical buttstroke.

> *Note:* Interruption is defined as your weapon being entangled or knocked offline toward your left side.

- Return to the modified basic warrior stance by moving forward and follow-on with a slash and a straight thrust. Refer to Appendix A for appropriate safeties.

See figure 4-2.

Figure 4-2. Buttstroke Offline.

Section II
Side Choke

The purpose of a choke is to render your aggressor unconscious or gain control of a close combat situation through less than lethal force. Chokes are performed by either closing off the airway to the lungs, thereby preventing oxygen from reaching the heart, or by cutting off the blood flow to the brain.

When executed properly, a side choke, which is a blood choke, takes between 8 to 13 seconds for the aggressor to lose consciousness. The air choke is least preferred because it takes longer to render the aggressor.

The side choke is a blood choke that is particularly effective when deflecting a punch thrown by an aggressor. It is performed when you and your aggressor are facing each other. When training, do not have the students execute the entire technique, instead walk the students step by step, focusing on arm placement and execution.

Refer to appendix A for corresponding safeties 1, 2, 6, 10, 11, and 12.

Technique
- Begin by facing the aggressor.
- The aggressor executes a rear hand punch. Block the aggressor's arm inboard with your lead forearm.
- Bring your right arm from underneath the aggressor's right arm and place your wrist or radius bone across your aggressor's carotid artery along the left side of the neck. Your right palm should be face down with your fingers extended and your thumb pointing toward you.
- With your left hand, reach around the back of the aggressor's neck and clasp your hands together with the left hand, palm side up. The aggressor's right arm should be over your right shoulder.
- Pull the aggressor toward your chest, exerting pressure on his left carotid artery with your right radius bone. This is accomplished by pulling your clasped hands toward your chest. At the same time, push up with your shoulder and head against the aggressor's triceps, driving his right shoulder into his right carotid artery. This allows you to apply pressure to both sides of the neck and not the trachea or windpipe.
  - Your shoulder and head should be placed high on the triceps close to the armpit to ensure that the shoulder is being driven in more effectively. Your aggressor's side should be up against you.
  - Ensure that pressure is exerted against either side of the neck and not on the throat (trachea or windpipe).

See figure 4-3 on page 4-8.

Figure 4-3. Side Choke.

Section III
Shoulder Throw

The purpose of a shoulder throw is to bring an aggressor to the ground and gain the tactical advantage in a fight. If an aggressor is moving toward you to attack, a shoulder throw can be used to take him to the ground while you remain standing. A shoulder throw is particularly effective if the aggressor is moving forward or pushing on you. Execution of the shoulder throw uses the aggressor's forward momentum.

There are a minimum of 10 fit ins for each throw during sustaining. Walk through the technique, step by step, working on proper body positioning and execution.

Refer to appendix A for corresponding safeties 1, 2, 6, 13, and 15.

Fit in
- Stand facing the aggressor in the basic warrior stance.
- Grab the aggressor's right wrist with your left hand and pull it into your left hip.
- Step forward with your right foot to the inside of the aggressor's right foot. Your heel should be between the aggressor's feet and your toes should be even with the aggressor's toes.
- Step back with your left foot, rotating on the ball of your right foot. Your heels should come close together as if you were in the position of attention. Your feet should be in between the aggressor's feet with your knees bent.
- At the same time, under-hook the aggressor's right arm with your right arm, pinching his arm between your biceps and forearm. You may grasp the aggressor's upper arm with your right hand for more control.
  - Hand placement should allow you to control the aggressor and pull him in close to you.
  - Your backside should be up against the aggressor. Your hips should be slightly lower than your aggressor's hips.
- Pull the aggressor's arm across your body. Lift the aggressor by straightening your legs and bending slightly at the waist.

CAUTION
Have students stop prior to throwing the aggressor to the deck. Practice this step prior to continuing.

~ Pull your aggressor's arm down and away with your left hand while bending straight over at the waist; throw your aggressor over your right shoulder.
~ Return to the basic warrior stance.

See figure 4-4.

Figure 4-4. Shoulder Throw.

Section IV
Counter to Strikes

A counter to a strike counters the aggressor's attack and allows the Marine to gain the tactical advantage.

Refer to appendix A for corresponding safeties 1, 2, 3, 6, 10, 11, 12, 13, and 15.

Counter to a Round Punch

The counter to a round punch will stop your aggressor's forward attack and gain control of the situation.

Technique

- From the basic warrior stance, begin with the aggressor extending his right arm as if executing a round punch, forward-left to the inside of the aggressor's attacking arm.
- This immediately moves your body out of the line of attack and places you inside the strike. Do not move backwards. You increase your chances of being hit because the outside of your aggressor's arm is moving faster than the inside and this is where all the power is generated.
- Attack with both arms bent so that your forearms make contact with the aggressor's biceps and forearm. Block the attack with the force of a strike, applying the principle that every block is a strike.
- With your left arm, over hook the aggressor's right arm at or slightly above the elbow.

~ Control the aggressor's arm by pinching it between your bicep and torso and execute a right inside knife hand strike to the right side of your aggressor's neck.

~ Grab the back of your aggressor's neck with the right hand, apply downward pressure, execute a right vertical knee strike to the aggressor's available target areas. After a minimum of three combination strikes, create distance between you and your aggressor. Return to the basic warrior stance.

See figure 4-5.

Figure 4-5. Counter to a Round Punch.

Counter to a Round Kick

The counter to the round kick is used when the aggressor executes a round kick with his right leg. A counter to a round kick allows you to take your aggressor to the deck and gain the tactical advantage over the situation.

Technique

- Begin with the aggressor extending his right leg, approximately waist level or slightly higher, in a round kick fashion. As the student's proficiency increases, the technique can be taught so that the student can defend against an aggressor executing a round kick.
- From the basic warrior stance, forward-left to the inside of the aggressor's attacking leg.
- This moves your body out of the line of attack and places you inside the power of the strike.
- At the same time, block your aggressor's attacking leg with the meaty portion of your forearms. Do not bend down to block the attack. Make two points of contacts on the attacking leg with both of your arms.
- Wrap your left arm over the aggressor's attacking leg, at or below the knee, and trap it between your bicep and torso.
- With your right hand, forcefully grasp the aggressor's face, push forward and to the left to further off balance the aggressor.

- For greatest effectiveness and efficiency of movement, you would insert your fingers deeply into the aggressor's eyes as you grasp his face. Executing a chin jab is acceptable.
- Simultaneously execute a leg sweep and drive the aggressor to the deck by pushing with your right arm against your aggressor's upper torso.

See figure 4-6.

Figure 4-6. Counter to Round Kick.

Section V
Lower Body Strikes

Lower body strikes are used to stop an aggressor's attack or create an opening in his defense in order to launch an attack. The legs provide the most powerful weapons of the body with which to execute strikes because they use the largest muscles of the body and are less prone to injury. The feet are the preferred choice for striking because boots protect them. The feet, heels, and knees of the legs are used to execute knee strikes, kicks, and stomps. Never kick high, because this jeopardizes your balance and leaves you more vulnerable to a counterattack.

The push kick is executed when the aggressor is in front of you and you need to stop an aggressor's attack or to create an opening in his defense in order to launch an attack. The striking surface is the ball of the foot. Primary target areas of the body are the aggressor's upper torso, any targets below the waist line, and the front of the thighs all the way down to the knees.

Technique
~ Always keep your right knee waist high or parallel to the deck.
~ Always keep your hands up to protect yourself from any strikes.
~ Make contact on the aggressor's abdomen with the ball of your rear foot. There is limited movement on one leg, so keep in mind that it is difficult to change the direction of a kick after it is initiated.
~ Follow through the target area by thrusting your hips forward toward your aggressor.
~ Rapidly return to the basic warrior stance.

See figure 4-7 on page 4-16.

Figure 4-7. Push Kick.

Section VI
Unarmed Manipulations

The purpose of unarmed manipulations is to teach Marines how to operate within the continuum of force, particularly in support of peacekeeping- or humanitarian-type missions. In these situations, Marines must act responsibly to handle situations without resorting to deadly force. Unarmed restraints and manipulation techniques such as enhanced pain compliance, the reverse wristlock come-along, and controlling techniques can be used to control an aggressor.

Refer to appendix A for corresponding safeties 1, 2, 6, 8, 9, 10, and 13.

Enhanced Pain Compliance

Applying two points of pain enhances control and leverage while applying a technique.

Technique
~ From a reverse wristlock, continue to control the aggressor's right hand with your right hand. Use your left hand to pull down and apply pressure on the aggressor's radial nerve. When pressure is added to the radial nerve, do not loosen your right hand's grip on the reverse wristlock. See figure 4-8 on page 4-18.
- Pressure can be added to the nerve by moving out and down, in a circular motion.
- It is acceptable to strike with the left hand before grabbing the radial nerve is.

4-18   MCRP 3-02B

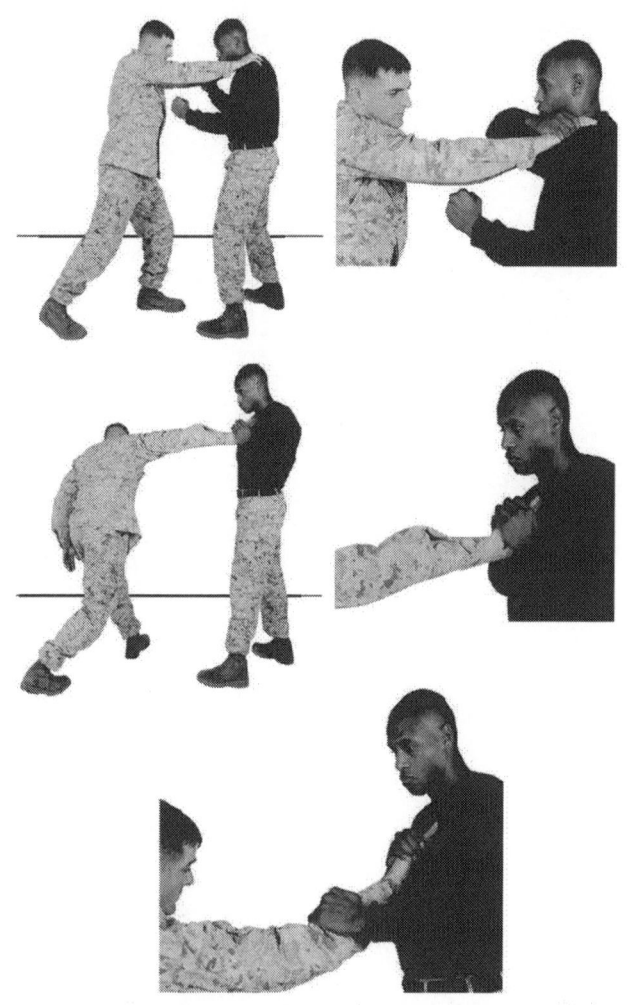

Figure 4-8. Enhanced Pain Compliance From a Reverse Wristlock.

~ From the basic wristlock, maintain control of the meaty portion of the aggressor's left hand with your right hand, use the left hand to maintain control of the lower forearm and elbow with the palm side up. Taking the left thumb and pushing it into the aggressor's ulnar nerve located just above the joint will add pressure. Pulling up on the elbow in the opposite direction of the hand also adds pressure. See figure 4-9.

For Official Use Only   Chapter 4: Green Be

Marine Corps Martial Arts Program 4-19

Figure 4-9. Enhanced Pain Compliance from a Basic Wristlock.

~ From the wristlock come-along, maintain control of the wrist with both hands, and lock the aggressor's elbow with the forearms. Apply pressure against the finger joints to bend them away from each other, splitting the fingers in opposite directions. See figure 4-10.

Figure 4-10. Enhanced Pain Compliance from a Wristlock Come-Along.

Reverse Wristlock Come-Along

The reverse wristlock come-along is effective when transporting a subject from one location to another. It can be used as follow-on technique to the reverse wristlock.

Technique
- ~ With your right hand, execute a reverse wristlock. Instead of stopping the rotation at 90-degrees, continue the rotation until the aggressor's palm is facing straight up toward the sky.
- ~ Ensure you keep the aggressor's hand close to your chest.
- ~ With your left hand, grab the meaty portion of the aggressor's thumb and as much of the palm as possible.
- ~ With your right hand, collapse the aggressor's right arm by applying pressure or striking the inner portion of his elbow.
- ~ At the same time step forward with the left foot pivoting on the ball of your right foot so that you are facing in the same direction as your aggressor.
- ~ With your right hand pull the aggressor's elbow to your chest as you bring your left arm high into your aggressor's armpit.
- ~ Maintain upward pressure with your left arm to keep your aggressor off balance.
- ~ Keep pressure on your aggressor's wrist by rotating your palm toward you while keeping downward pressure on the hand to maintain compliance over your aggressor.
- ~ Your right hand will move on top of the aggressor's right hand. Keep your elbows tight to control the aggressor's arm.

Marine Corps Martial Arts Program 4-21

See figure 4-11.

> *Note:* For enhanced pain compliance release with your right hand and re-grasp the bottom two fingers on the aggressor's hand and pull them down and away from his hand.

Figure 4-11. Reverse Wristlock Come-Along.

Chapter 4: Green Belt For Official Use Only

Controlling Techniques

Opposite Side Grab

The opposite side grab is effective if someone tries to grab your wrist.

Technique
- Begin with the aggressor grabbing your right wrist with his right hand.
- Trap the aggressor's right hand in place on the right wrist with the palm of your left hand.
- Rotate your right hand to reach up and grasp the aggressor's right forearm while maintaining downward pressure on the aggressors trapped right hand, with your left hand.
- The target area to grasp on the right forearm is the radial nerve.
- Apply downward pressure in and down with both hands until the aggressor is forced to one knee and is effectively controlled.

See figure 4-12.

Marine Corps Martial Arts Program 4-23

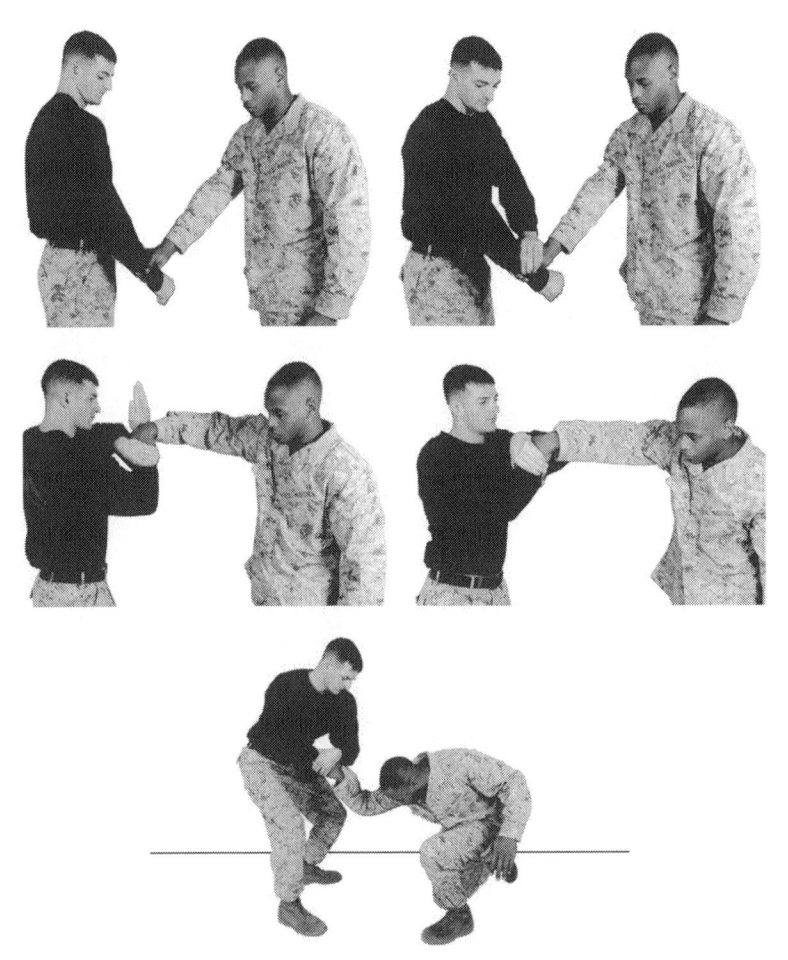

Figure 4-12. Opposite Side Grab.

Chapter 4: Green Belt — For Official Use Only

Same Side Grab

The same side grab is effective if someone tries to grab your wrist.

Technique
~ Begin with the aggressor grabbing your right wrist with his left hand.
~ Rotate your right palm upward.
~ Grab the backside of the aggressor's hand with your left hand, palm side up, wrapping your fingers around the meaty portion of his thumb.
~ Continue to rotate the aggressor's palm outboard until control or compliance is achieved.

See figure 4-13.

Figure 4-13. Same Side Grab.

Section VII
Knife Techniques

The purpose of knife techniques is to cause enough damage and massive trauma to stop an aggressor by properly executing a counter knife strike from a vertical attack and a forward strike.

Refer to appendix A for corresponding safeties 1, 2, 6, 8, 9, 10, and 13.

Counter to a Vertical Strike With Follow-on Techniques

Technique
- From the modified basic warrior stance, move forward, inside the arc of attack.
- Block the attack, over and in front of your head, with your left arm. The arm is bent so that your forearm makes contact with the aggressor's forearm.
- At the same time execute a vertical thrust into the aggressor's neck and follow through with at least three more killing techniques, preferably thrusting to available target areas. Control the aggressor's attacking arm throughout.

See figure 4-14.

Figure 4-14. Counter to a Vertical Strike.

Counter to a Forward Strike With Follow-on Techniques

Technique
- From the modified basic warrior stance, move forward-left, inside the arc of the attack.
- Block the attack with your left arm bent so that your forearm makes contact with the aggressor's forearm.
- At the same time, use the blade to block the aggressor's upper arm or bicep. Then slash downward on the biceps dropping your body weight to develop maximum force.
- Follow through with at least three more killing techniques preferably thrusting to available target areas on the aggressor. Control the aggressor's attacking arm throughout.

See figure 4-15.

Figure 4-15. Counter to a Forward Strike.

Section VIII
Weapons of Opportunity

The purpose of weapons of opportunity is to enable each individual Marine to be creative and utilize any object on the battlefield in order to inflict maximum damage to the aggressor. A Marine should be ready and able to use anything around him to serve as a weapon. This may mean throwing sand or liquid in an aggressor's eyes to temporarily impair his vision or executing blocks: the block for a vertical strike, the block for a forward strike, the block for a reverse strike, and the block for a straight thrust. Weapons of opportunity blocks are effective when you are blocking attacks and the aggressor is using a weapon of opportunity. In any given confrontation, a Marine must use whatever it takes to win and move on to the next aggressor.

Refer to appendix A for corresponding safeties 1, 2, 3, 4, and 14.

Block for a Vertical Strike With Follow-on Strikes

All blocks being taught in this lesson will begin from the one-handed carry or two-handed carry. When the aggressor has a weapon, blocks are executed with your weapon.

Technique
- Begin in the modified basic warrior stance with a one-handed grip. The aggressor extends his right hand in a vertical attack.
- Move forward to get your body inside the arc of attack. This movement gets you inside the aggressor's strike and his generated power. When you step in forcefully, you prevent the aggressor from developing power in his swing and you stop his momentum.
- Block the aggressor's weapon by making two points of contact to disperse the impact of the attack.
- Block the aggressor's weapon by positioning your weapon so it is perpendicular to the aggressor's weapon. If your weapon is not perpendicular to the aggressor's weapon, the aggressor's weapon can slide through and make contact on you.
- With the muscular portion of your left forearm, block the aggressor's wrist or forearm with two points of contact. Control the arm with your left arm or grip the arm with your hand to maintain control.
- If you are closer to the aggressor, use your weapon to block the aggressor's arm. It is the same movement, except now you block the aggressor's arm with both your weapon and your arm.
- Follow up with strikes to lethal target areas on the aggressor.

See figure 4-16 on page 4-28.

4-28   MCRP 3-02B

Figure 4-16. Block for a Vertical Strike.

Block for a Forward Strike With Follow-on Strikes

═══ Technique

~ Begin in the modified basic warrior stance with a one-handed carry. The aggressor extends his right hand in a forward strike.
~ Move forward-left to get your body inside the arc of the attack.
~ To disperse the impact of the attack, block with two points of contact. Block the aggressor's wrist or forearm with the meaty portion of your left forearm and maintain control of the arm.
~ Strike the aggressor's attacking biceps with your weapon.

For Official Use Only   Chapter 4: Green Be

Marine Corps Martial Arts Program 4-29

~ Control the arm with your left arm or grip the arm with your hand to maintain control.
~ Follow up with strikes to lethal target areas.
See figure 4-17.

Figure 4-17. Block for a Forward Strike.

Block for a Reverse Strike With Follow-on Strikes

=== Technique
~ Begin in the modified basic warrior stance with a one-handed carry. The aggressor extends his right hand in a reverse strike.
~ Move forward-right to get your body inside the arc of the attack.

Chapter 4: Green Belt For Official Use Only

- To disperse the impact of the attack, block with two points of contact. Block the aggressor's weapon by positioning your weapon so that it is perpendicular and making contact with the aggressor's weapon.
- With your left arm, block the aggressor's forearm with the muscular portion of your forearm. If you are in closer to the aggressor, block the aggressor's triceps with the back of your left forearm and strike his forearm with your weapon.
- Control the arm with your left arm or grip the arm with your hand to maintain control.
- Follow up with strikes to lethal target areas.

See figure 4-18.

Figure 4-18. Block for a Reverse Strike.

Block for a Straight Thrust With Follow-on Strikes

Technique

- Begin by facing the aggressor with his right arm straight out in front of him, simulating a straight thrust.
- Block your aggressor's attack by striking his arm perpendicular with yours as you step forward-right with your right foot, to the outside of your aggressor's left foot.
- With your left hand, reach across the block and grasp and control your aggressor's attacking limb. This will prevent a follow-on attack with his weapon and leave him open for you.
- Follow up with strikes to lethal target areas.

See figure 4-19.

Marine Corps Martial Arts Program 4-31

Figure 4-19. Block for a Straight Thrust with Follow-on Strikes.

Chapter 4: Green Belt   For Official Use Only

Section IX
Ground Fighting

In any close combat situation, the fight may end up on the ground. The purpose of ground fighting techniques is to allow you to return to your feet as quickly as possible and regain the tactical advantage. Techniques that can be used in ground fighting are the armbar from the mount position and the armbar position from the guard position.

Refer to appendix A for corresponding safeties 1, 2, 6, and 10.

Armbar From Mount Position

The armbar from the mount position is effective for causing damage to an aggressor's arm providing you with a tactical advantage and allowing you to get on your feet.

Technique

- Begin with the aggressor on his back, on the ground.
- Sit astride on the aggressor's abdomen. Both legs are bent, with your knees touching the ground. Use your weight and hips to control your aggressor. Keeping all of your weight on your knees makes it easier for the aggressor to maneuver and escape.
- The aggressor attempts to choke or push you off of him with straight arms.
- Place both palms in the center of the aggressor's chest. Your right arm will weave over the aggressor's left arm and your left arm under his right arm. Post the majority of your weight on your two hands pinning the aggressor to the ground.
- Keep your right hip/upper, inside thigh in constant contact with your aggressor's body, creating pressure on the aggressor's left triceps. Turn your body to face left and swing your right leg over the aggressor's head.
- Pull your feet in toward your buttocks to trap his body and pinch your knees together to trap his arm, hug his left arm to your chest and sit back with your upper body to straighten and lock out the aggressor's arm.
- Maintain pressure against the aggressor's neck with the back of your right foot and against his side with your left foot underneath his armpit. Your legs are on either side of the aggressor's left arm. Use your hips to make pressure into the aggressor's elbow in the direction of his left pinky. Your aggressor's left palm should be facing the sky.
- Falling back quickly and thrusting your hips up strongly against the aggressor's elbow can break or dislocate the aggressor's elbow. It is important to do this slowly in training in order to give your training partner an opportunity to tap out and avoid injury. The aggressor will tap out by tapping on the deck three times, on himself three times, or by verbally saying *tap-tap-tap*.
- Return to your feet, gaining the tactical advantage. You can follow-on with controlling techniques to lethal force depending on the temper and intent of your adversary.

See figure 4-20.

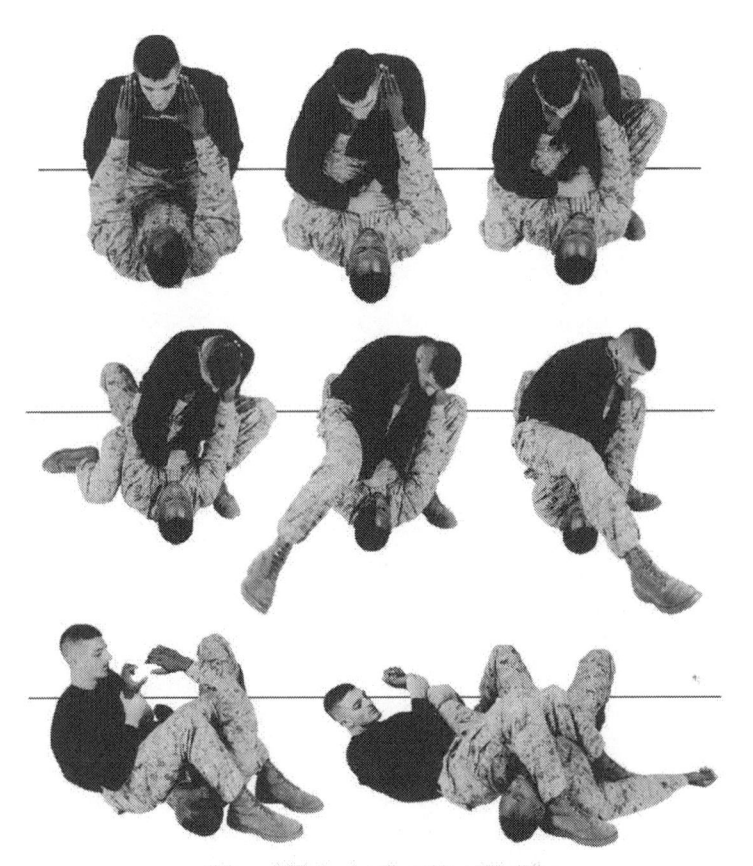

Figure 4-20. Armbar From Mount Position.

Armbar From Guard Position

The armbar from the guard position is effective for causing damage to an aggressors arm and providing you with a tactical advantage and allowing you to get on your feet.

Technique

- Begin by lying on your back with the aggressor kneeling between your legs with his hands around your neck. Wrap your legs around the aggressor's waist.
- Trap the aggressor's hands on your chest by crossing your hands on your chest so that your forearms are resting on his/her forearms.
- With the cutting edge of your right heel, strike the outside of the aggressor's left thigh. This will cause the aggressor to jerk to that side.
- Quickly move your head to your left and swivel your hips to your right while maintaining positive control of your aggressor's left arm. Simultaneously, bring up both of your legs so that they are on the right side of the aggressor's body.
- Bring your right leg down, hooking the aggressor's neck and head, and exert downward pressure to roll him over on his back. Grasp and maintain control of the aggressor's left arm.
- You should end up sitting up with your legs bent over the aggressor while maintaining control of his left arm. Your buttocks are tight against the aggressor's shoulder. Move closer to your aggressor if necessary. Your legs are on either side of the aggressor's right arm.
- Keeping your legs and knees bent, maintain pressure against the aggressor's neck with the back of your right foot and your left foot underneath his armpit against his side. Squeeze your knees together, tightly locking in the aggressor's arm.
- Pull the aggressor's arm straight up and fall back sharply, pulling his arm to the side in the direction of his little finger. Your aggressor's left palm should be facing the sky. This action will break the aggressor's arm. Raise your hips slightly and pinch your toes in around the aggressor's shoulder to maintain control and to generate power in the break.
- Return to the basic warrior stance.

See figure 4-21.

Figure 4-21. Armbar From Guard Position.

THIS PAGE INTENTIONALLY LEFT BLANK.

CHAPTER 5

Brown Belt

The fourth belt ranking within MCMAP is Brown Belt. Upon qualifying as a Green Belt, all Marines are highly encouraged to continue MCMAP sustainment training in order to advance to Brown Belt. Brown Belt is the introduction to advanced fundamentals of each discipline. Brown Belt is the minimum training goal of all infantrymen. Purpose and principles remain the same as in Gray Belt.

Brown Belt Requirements.

Prerequisites	Recommendation of reporting senior
	Complete Green Belt sustainment and integration training
	Complete PME
Training Hours	A minimum of 18.5 hours, excluding remedial practice time and testing
Sustainment Hours	A minimum of 15 hours of sustainment, excluding integration training time and practice time for testing

For Official Use Only

Section I
Bayonet Techniques

The purpose of bayonet techniques is to disable or kill the aggressor. When engaging in combat, mindset, more often than not, will be the determining factor of success or failure, regardless of technical proficiency. Anyone can train in a martial skill, but few have the mind and will to use their skills to kill or injure. Mindset is often the mental trigger in the defining moment that forces you to commit to an aggressor with the goal of injury or death.

If faced with one-on-two engagements, two-on-one engagements, or two-on-two engagements, execute the bayonet techniques such as the straight thrust and the slash that you learned earlier in your martial arts training.

The following bayonet training principles are applied to bayonet fighting:

Disrupt. A disrupt creates an opening in the aggressor's defense by bringing the aggressor's weapon off line.

Entry. Entry is the movement that is made in order to get inside the aggressor's defense and find a path to the target. A movement can be a step forward or a small step to an oblique to get within striking distance of the aggressor.

Modified Basic Warrior Stance. All movement begins and ends with the basic warrior stance.

Channeling the Aggressor. Move using the approach and close principles of movement while attempting to place one aggressor in front of the other. Use the angles of approach learned earlier in your martial arts training.

Refer to appendix A for corresponding safeties 1, 2, 4, 7, 8, 9, and 14.

One-on-Two Engagement

Technique
~ Assume the modified basic warrior stance.
~ Stagger two aggressors, approximately 10 meters in stationary positions, away from you. See figure 5-1.

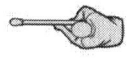

Figure 5-1. One-on-Two Engagement.

Marine Corps Martial Arts Program

- ~ Execute movement toward aggressor.
- ~ While moving toward aggressor, use angles of approach to place one aggressor in front of the other.

See figure 5-2.

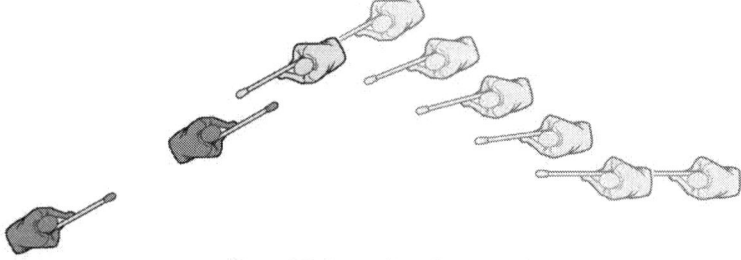

Figure 5-2. One-on-Two Engagement.

- ~ Use an economy of motion and no hesitation while moving toward aggressors.
- ~ Utilize natural terrain features and/or obstacles to channel aggressors along with movement.
- ~ Attempt to engage one aggressor at a time. Do not hesitate at any time.
- ~ The primary technique used is the straight thrust, unless a disrupt is needed to disrupt your aggressor.
- ~ When at closing distance, the student should begin to thrust bayonet trainer so that the students can engage the aggressor at the furthest possible distance.

Two-on-One Engagement

==== Technique
- ~ Assume the modified basic warrior stance.
- ~ Place two Marines, in a stationary position, 10 to 20 meters away from the aggressor.

See figure 5-3.

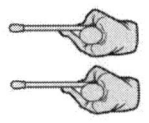

Figure 5-3. Two-on-One Engagement.

- ~ Execute movement toward the aggressor.
- ~ While moving toward the aggressor, do not allow him to use angles of approach to place one Marine in front of the other.

~ Stay together in a manner that allows Marines to attack simultaneously, covering each other's movement and attack. This can be accomplished by either staying shoulder-to-shoulder or within a 90-degree angle of approach to each other. Communication with each other is essential.

See figure 5-4.

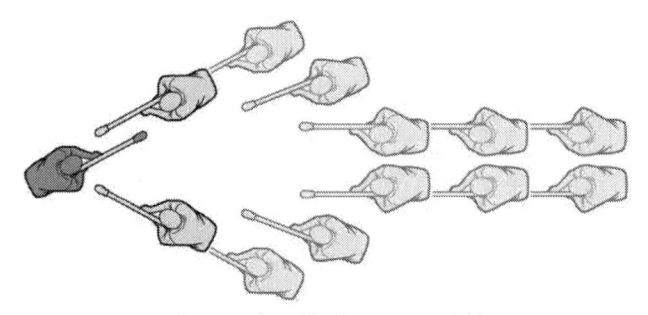

Figure 5-4. Two-on-One Engagement.

~ Use an economy of motion and no hesitation while moving toward the aggressor.
~ Utilize natural terrain features and/or obstacles to canalize the aggressor along with movement.
~ The primary technique used is the straight thrust, unless a disrupt is needed to clear a path for your blade.

Two-on-Two Engagement

Technique

~ Two Marines assume the modified basic warrior stance.
~ Stagger two aggressors in a stationary position, 10 to 20 meters away from the other Marines.

See figure 5-5.

~ Execute movement toward the aggressors.
~ While moving toward the aggressors, use angles of approach to place one aggressor in front of the other. At the same time, do not allow the aggressors to use angles of approach and movement to place you in front of the other Marine.

Marine Corps Martial Arts Program 5-5

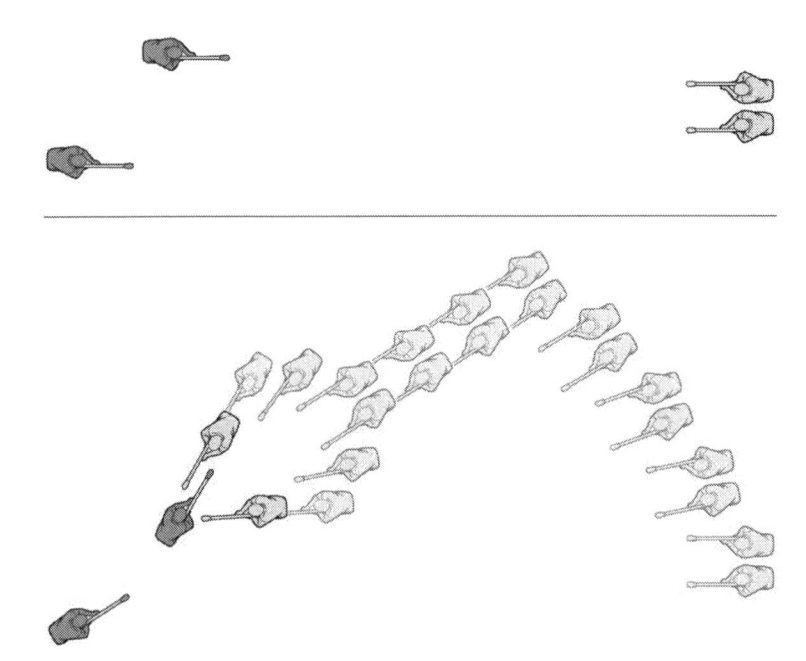

Figure 5-5. Two-on-Two Engagement.

~ Stay together in a manner that allows both Marines to attack one aggressor at a time, covering each other's movement and attack. This can be accomplished by either staying shoulder-to-shoulder or within a 90-degree angle of approach to each other. Do not allow the aggressors to turn this into two, one-on-one engagements.

~ When at closing distance, begin to thrust the bayonet so that you can engage the aggressor at the furthest possible distance.

Chapter 5: Brown Belt For Official Use Only

Section II
Ground Fighting

The purpose of ground fighting is to apply techniques that will allow you to get to your feet as quickly as possible and cause damage to the aggressor.

Refer to appendix A for corresponding safeties 1, 2, 6, and 10.

Ground Fighting Techniques

Ground fighting techniques are used to cause damage to the aggressor using the bent armbar from the side mount position and the basic leg lock, which will enable you to get to your feet.

Bent Armbar From the Side Mount Position

The bent armbar from the side mount position technique can be executed from many positions, but we will focus on executing this technique from the side mount position, while the aggressor uses his free hand to attempt an eye gouge.

Technique
- Begin with the aggressor lying flat on his back. Position yourself in the right side mount position, chest to chest with the aggressor.
- Your left leg is straight and your right leg is bent, on the ground at the aggressor's hip. Your body weight will control the aggressor.
- The aggressor attempts an eye gouge with his left hand.
- With your left hand, grab the aggressor's left wrist and place your left elbow against the aggressor's left ear.
- With your right hand, reach underneath the aggressor's left triceps and grab your left wrist, both palms down.
- Apply downward pressure as you pull the aggressor's wrist toward his hips or feet, simultaneously elevating his left elbow by raising your right elbow. For submission, apply slow and steady pressure.
- For joint destruction, rapidly execute full force and full speed while raising the right elbow or your aggressor's left elbow.
- Return to your feet and assume the basic warrior stance.

See figure 5-6.

Marine Corps Martial Arts Program

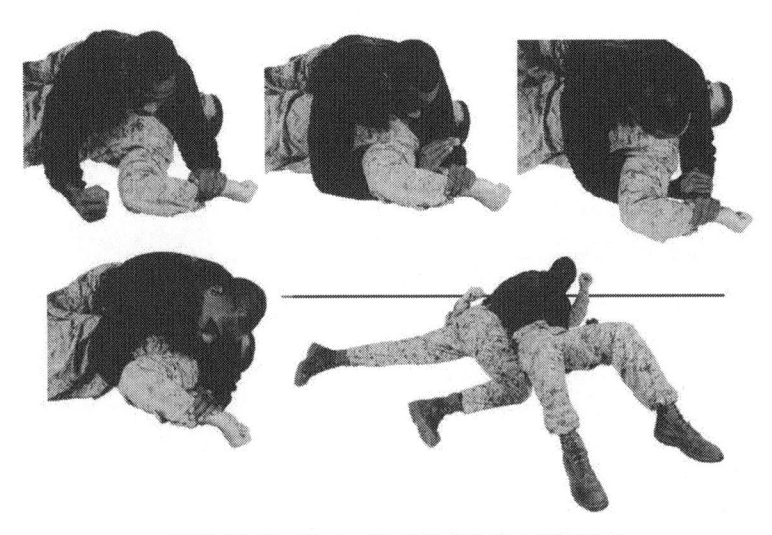

Figure 5-6. Bent Armbar From the Side Mount Position.

Basic Leg Lock

The basic leg lock can be executed when you are in your aggressor's guard, prior to him locking his ankles together. In the following scenario, the aggressor is attempting to gain the superior position.

Technique
~ Begin in the aggressor's guard without him locking his ankles together.
~ With your left elbow, strike the inside of the aggressor's right leg, on the femoral nerve. Maintain pressure on his leg, pinning it to the ground with your left hand until you replace your hand with your left foot in step five.
~ At the same time bring your right leg forward and to the right, at a 45-degree angle, placing the sole of the foot on the deck.
~ Quickly over hook your right arm around the aggressor's left leg, between the lower calf and the heel.

- Quickly come to the standing position, maintaining control of the aggressor's legs, keeping your back straight and your knees slightly bent. With your left leg, execute a vertical stomp on the aggressor's groin region. For safety, place your foot on your aggressor's thigh.

CAUTION
For safety during training, simulate stomping the groin by placing your left foot on the inside of the aggressor's thigh.

- Clasp your hands together; palm-to-palm, with your right palm toward the deck. Apply bone pressure to the aggressor's Achilles tendon by rotating your right radius bone up and into the Achilles tendon. Maintain pressure on the aggressor's Achilles tendon throughout the movement. Slowly straighten your back and look up to increase the pressure for the submission. Stand quickly, arching your back for the joint destruction.
- Return to the basic warrior stance.

See figure 5-7.

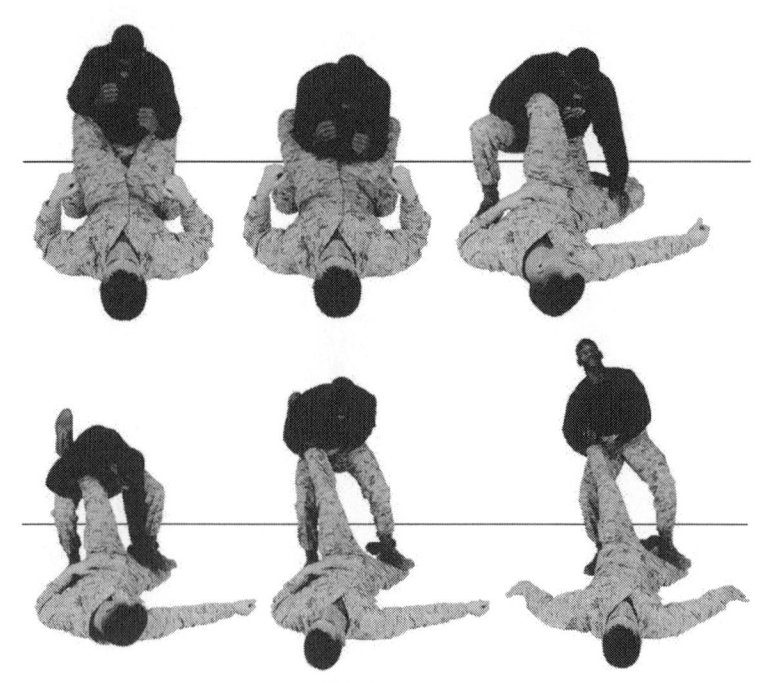

Figure 5-7. Basic Leg Lock.

Section III
Ground Chokes

The purpose of ground chokes is to render the aggressor unconscious as quickly as possible by using the rear ground choke, figure-4 variation, side choke, or front choke.

Refer to appendix A for corresponding safeties 1, 2, 6, 10, 11, and 12.

Rear Ground Choke

The rear ground choke is a blood choke that is performed when you are behind the aggressor on the ground.

Technique
- Begin by sitting on the ground with the aggressor sitting between your legs with his back to your chest.
- Place your lower legs over the aggressor's thighs, this should be done simultaneously or after the choke is executed in order to prevent a counter. Do not cross your ankles at any time, this will compromise your position and places you in a vulnerable position to counter-attack. Use your insteps and toes to create constant pressure on the aggressor's thighs.

  Note: Do not try to get your toes under the aggressor's legs.

- With your left arm, reach up and grab the aggressor's forehead and pull back.
- With your right arm, reach over the aggressor's right shoulder and hook the bend between the forearm and bicep of your arm around his neck. Ensure that the aggressor's windpipe is positioned within the bend of your arm and that pressure is not being exerted directly on his windpipe.
- With your left hand, palm side up, clasp both hands together, palm-to-palm.

~ Exert pressure with your biceps and forearm on the carotid arteries on both sides of the aggressor's neck; while maintaining pressure, draw the aggressor closer to you by drawing your right arm in.
~ To increase the effectiveness of the choke, apply forward pressure to the back of the aggressor's head with your head by bending your neck forward. Stretch out the aggressor by straightening and arching your body while maintaining your hooks and arm positions.

See figure 5-8.

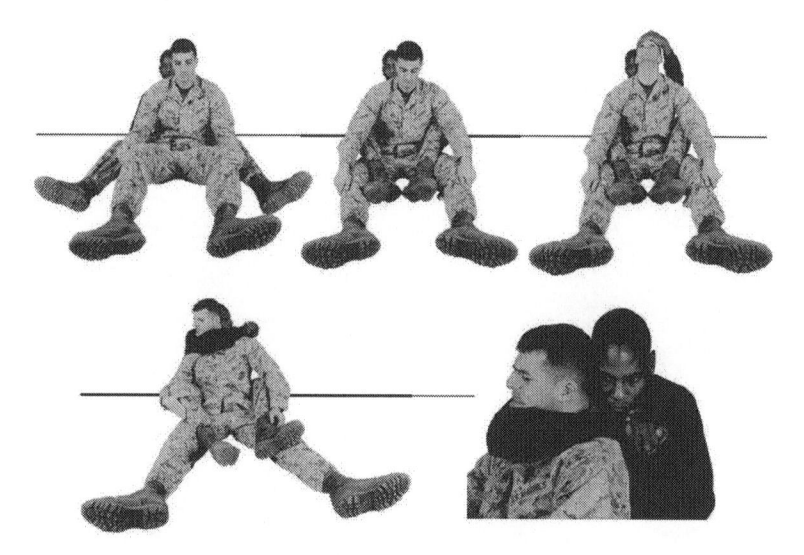

Figure 5-8. Rear Ground Choke.

Figure-4 Variation of Rear Choke

The figure-4 variation of the rear choke is also a blood choke and is performed when you are behind the aggressor.

Technique
~ Follow steps one through three for the rear choke.
~ Grasp your left bicep or shoulder with your right hand and place your left hand against the back of the aggressor's head.
~ With your left hand, push the aggressor's head forward and down.
~ Draw your right arm in, maintaining pressure with your biceps and forearm on both sides of the aggressor's neck.
~ To increase the effectiveness of the choke, stretch the aggressor out by straightening and arching your body while maintaining your leg and arm positions.

See figure 5-9.

Figure 5-9. Figure-4 Variation of Rear Choke.

Front Choke

The ground front choke is a blood choke performed when you are in your aggressor's guard or in the mount position. This choke is performed the same as outlined in chapter 3, Gray Belt.

Technique
~ Begin in the mount position.
~ With your right hand, grab the back of the aggressor's right collar, making certain that your palm is facing up.
~ Keeping the collar tight in your right palm, reach under your right arm with your left hand and grab the back of the aggressor's left collar, making certain that your palm is facing up, forming an X with your wrists. Attempt to make your thumbs touch.
~ Grab the collar with your elbows facing down, curl your wrist inward, and pull down while at the same time attempt to place your head over the aggressor's left shoulder. Your radius bone will cut off the aggressor's carotid artery.
~ Make sure you apply pressure on the carotid artery and not on the throat (trachea or windpipe).

See figure 5-10.

Figure 5-10. Front Choke.

Side Choke

The side choke is a blood choke performed from the mount position and is particularly effective when the aggressor raises his arms and places them on your chest or throat. The side choke is performed the same as outlined in the Green Belt chapter.

Technique
- Begin facing the aggressor in the mount position.
- The aggressor will push up on you in an effort to get you up and off of him.
- Disrupt the aggressor's right arm inboard, cross-chest with your left hand.
- Bring your right arm underneath the aggressor's right arm and place your wrist and radius bone across the aggressor's carotid artery along the left side of his neck. Your right palm should face down with your fingers extended and your thumb pointing toward you.
- With your left hand, reach around the back of the aggressor's neck and clasp your palms together with your left hand, palm side up. The aggressor's right arm should be over your right shoulder.
- Pull the aggressor toward your chest, exerting pressure on the side of his neck with your forearm. At the same time, push up with your shoulder and head against the aggressor's triceps, driving his shoulder into his right carotid artery. This allows you to apply pressure on the carotid artery and not the trachea or windpipe. Finish by pulling your clasped hands towards your chest.

See figure 5-11.

Marine Corps Martial Arts Program

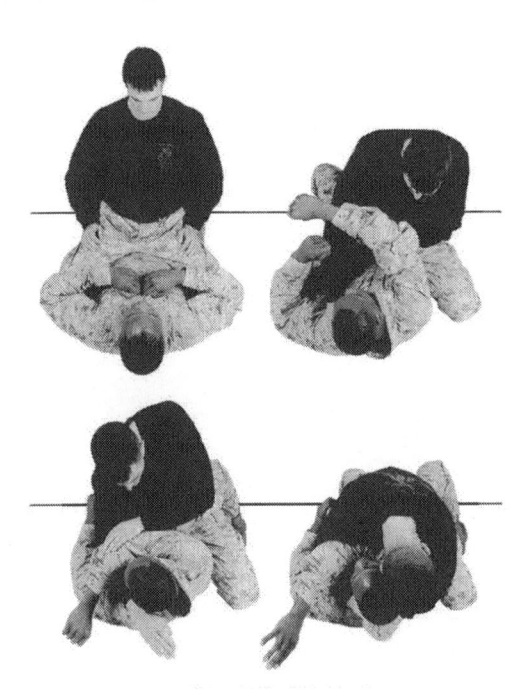

Figure 5-11. Side Choke.

Section IV
Major Outside Reap Throw

This technique is used to bring your aggressor to the ground from an extremely close range when tied up in a clinch using the outside reap throw, both pushing and pulling.

Refer to appendix A for corresponding safeties 1, 2, 3, 6, 10, 11, 12, 13, and 15.

Major Outside Reap Throw: Aggressor Pushing

A major outside reap throw can be used to take the aggressor to the deck while you remain standing. It is particularly effective if the aggressor is already off balance, pushing you and defending against the leg sweep.

Technique
- From the clinch, (right hand behind the aggressor's neck and left hand on aggressor's right triceps), off balance the aggressor by pulling his right arm in and downward with your left hand.
- With your right hand on the left side of his neck or head, push down and to your left in an arcing movement.
- While off balancing the aggressor, just before weight is transferred back on to the aggressor's right foot, sweep his right leg across his body with the inside of your left foot.

See figure 5-12.

CAUTION
For safety during training, make sure that you do not strike directly on your aggressor's ankle with the inside of your boot.

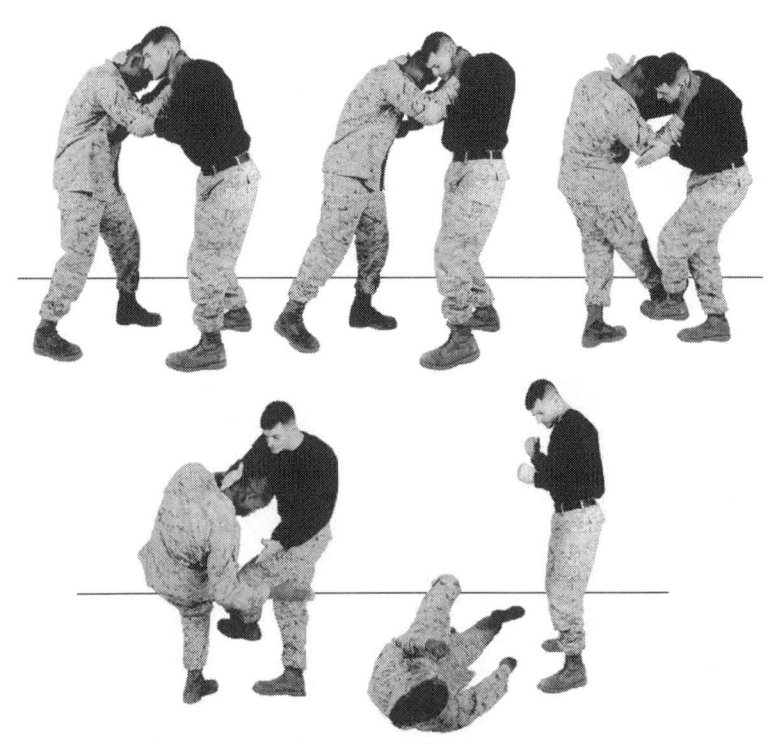

Figure 5-12. Major Outside Reap Throw: Aggressor Pushing.

Major Outside Reap Throw: Aggressor Pulling

A major outside reap can be used to take the aggressor to the deck while you remain standing. It is particularly effective if the aggressor is already off balance, pulling you and defending against the leg sweep.

Technique
- From the clinch (right hand behind the aggressor's neck and left hand on the aggressor's right triceps) off balance the aggressor by pulling his right arm in and downward with your left hand.
- With your right hand on the left side of the aggressor's neck or head, push down and to your left in an arcing movement.
- While off balancing the aggressor, just before weight is transferred off of the aggressor's right foot, sweep his right leg across his body with the inside of your left foot.

See figure 5-13.

CAUTION

For safety during training, make sure that you do not strike directly on your aggressor's ankle with the inside of your boot.

Marine Corps Martial Arts Program

Figure 5-13. Major Outside Reap Throw: Aggressor Pulling.

Section V
Unarmed Verses Handheld

The purpose of unarmed versus handheld is to be able to disarm the aggressor without the aid of a weapon using hollowing out and follow-on techniques, reverse armbar counter, or bent armbar counter.

In any engagement against a knife, a stick, or some other weapon of opportunity, you must establish and retain a mindset to go on the offensive rather than be on the defensive. Your survival depends on it.

A counter is used to control the situation, in order to regain the tactical advantage and end the fight. Regardless of the type of weapon or angle of attack, the following actions apply to countering the attack with a handheld weapon.

- The first action in a counter is to move out of the line of attack. Movement is executed in a 45-degree angle forward to the left or right.
- The second action is to block the attack.

   Note: The first and second actions are taken simultaneously.

- The third action is to control the weapon by controlling the aggressor's hand or arm, whichever is holding the weapon. Never attempt to grab the aggressor's weapon.
- The fourth action is to execute appropriate follow ups to end the fight such as strikes, joint manipulations, throws, or takedowns. You should continue your assault on the aggressor until you end the fight.

Refer to appendix A for corresponding safeties 1, 2 6, 10, and 13.

Hollowing Out with Follow-on Technique

Hollowing out allows you to move away from an aggressor's attack in order to create space for follow-on strikes.

==== Technique
~ Start from the basic warrior stance, facing the aggressor who attacks with a straight thrust.
~ Bend at the waist moving your hips backwards and jumping backwards with both feet moving away from the attack. This action is known as hollowing out.
~ Hollow out and block the attack with your elbows slightly bent and hands together on top of the attacking arm.
~ Your hands should be palm down and slightly overlapped so that one thumb is on top of the other hand's index finger; the other thumb should be under the other hand's index finger in an inverted V.
~ Maintain control of the aggressor's attacking arm by firmly grasping his wrist.
~ Follow-on techniques, such as strikes and joint-locks, are used to control the weapon, subdue the aggressor, and remove the weapon.

See figure 5-14.

Marine Corps Martial Arts Program 5-19

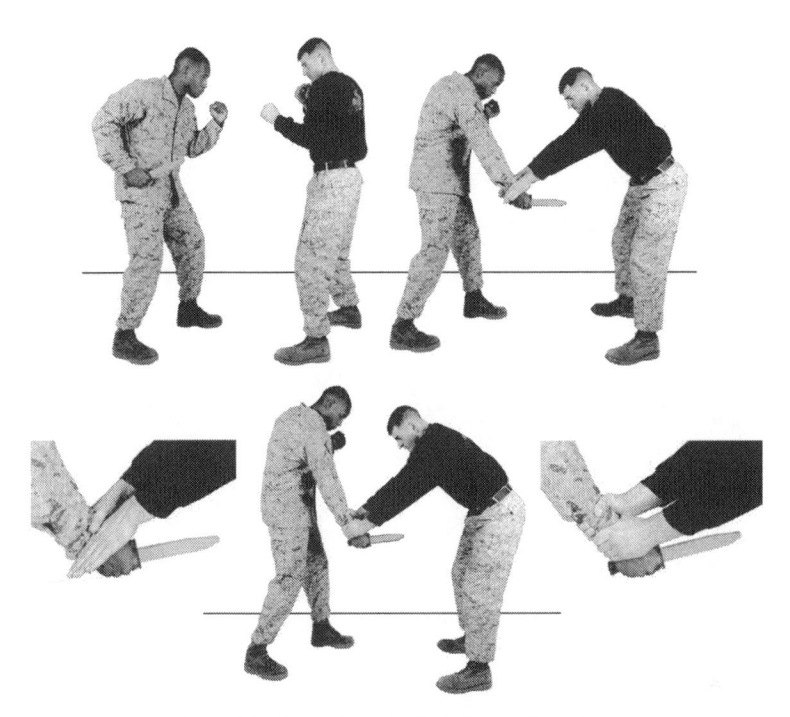

Figure 5-14. Hollowing out with Follow-on Technique.

Forward Armbar Counter

The forward armbar counter is effective against an opponent that is excuting a forward strike.

Technique
- Face the aggressor in the basic warrior stance. The aggressor attacks with a forward strike coming in anywhere from a 45-degree angle of attack to parallel to the deck.
- Move forward-left, inside the arc of the aggressor's attack.
- Block the attack with both arms bent so that the outside of your forearms makes two points of contact with the aggressor's biceps and forearm.
- Immediately after making two points of contact, over hook or wrap your left arm over the aggressor's arm, trapping his attacking arm between your bicep and your torso by pulling your elbow in. His forearm should be under your armpit with your left forearm making pressure on the aggressor's right elbow.
- Place your right hand on the aggressor's shoulder or upper arm and your left palm on your right wrist to further control his arm and to effect an armbar with your left forearm exerting pressure on his right elbow.

Chapter 5: Brown Belt   For Official Use Only

~ Follow-on techniques, such as strikes and joint-locks, are used to control the weapon, subdue the aggressor, and remove the weapon.

See figure 5-15.

Figure 5-15. Forward Armbar Counter.

Reverse Armbar Counter

The reverse armbar counter is effective aginst an opponent excuting a reverse strike.

Technique
- Face the aggressor in the basic warrior stance. The aggressor attacks with a reverse strike coming in anywhere from a 45-degree angle of attack to parallel to the ground.
- Move forward-right, outside the arc of the aggressor's attack.
- Block the attack with both arms bent so that your forearms make two points of contact with the aggressor's biceps and forearm.
- While maintaining control of the aggressor's wrist with your right hand, pivot to your right so that your back is against the aggressor's right side. Immediately over hook the aggressor's right arm with your left and wrap your arm tightly around his arm, trapping his attacking arm between your biceps and your torso.

- The aggressor's bicep should be under your armpit. You must control the aggressor's arm on his elbow in order to affect an armbar from this position.
- Control the aggressor's arm by pinching it between your arm and your torso.
- With your right hand, twist the aggressor's wrist and hand outboard, palm side up.
- Complete the armbar by grabbing the top of your right wrist with your left palm. (If your arms are too short it is permissible for you to grab your own gear or utilities at your chest with your left hand to make the pressure and secure the lock).
- Apply downward pressure on the aggressor's upper arm and shoulder. Your body is used to apply pressure by arching and/or dropping body weight into the aggressor.
- Apply upward pressure on the aggressor's elbow while applying downward pressure on his wrist.
- Perform follow-on techniques.

See figure 5-16.

Figure 5-16. Reverse Armbar Counter.

Bent Armbar Counter

The bent armbar counter is particularly effective against a vertical attack.

Technique

- Face the aggressor in a basic warrior stance. The aggressor attacks you with a vertical strike.
- Move forward-left to the inside of the aggressor's attacking arm.
- Block the attack with both arms bent so that your forearms make contact with the aggressor's biceps and forearm.
- With your left hand, grab the aggressor's right wrist. At the same time, slide your right arm underneath his triceps and hook his forearm or wrist with your right hand, hand over hand, bringing your elbows close together.
- Apply pressure forward and down with your hands against the aggressor's forearm to off balance him. Keep the aggressor's arm bent and elbow in close to your body to maintain leverage. This action can dislocate or damage the aggressor's shoulder.
- To take the aggressor to the ground, step past him with your right foot, while keeping the aggressor's arm tight into your body, execute the leg sweep.

See figure 5-17.

Figure 5-17. Bent Armbar Counter.

Section VI
Firearm Retention

Firearm retention techniques are designed to provide Marines with the skills necessary to maintain positive control of their weapons and, if necessary, restrain aggressors that are attempting to disarm them. This can be accomplished by executing the blocking technique, armbar technique, wristlock technique, or same side grabs.

Refer to appendix A for corresponding safeties 1, 2, 6, 8, 9, 10, and 13.

Blocking Technique

The blocking technique is used when the aggressor attempts to grab your pistol from the holster.

=== Technique
~ While you are facing the aggressor, he attempts to grab your holstered pistol with his right hand.
~ Step back with your right foot pivoting your body away from the aggressor while placing your hand on the grip of the pistol.
~ Extend the forearm of your left arm and block, deflect, or strike the aggressor's arm, while forcefully yelling *Get Back* or any authoritative command.
~ Continue to create distance between you and the aggressor to enable you to access and present your weapon or set up for follow-on actions that are appropriate to the situation and conditions.

See figure 5-18.

Figure 5-18. Blocking Technique.

Armbar Technique

The armbar technique is used when an aggressor grabs your pistol in the holster with his right hand or if you are left-handed and the aggressor grabs your pistol with his left hand.

Technique
- Begin with the aggressor facing you and grabbing the pistol in your holster with his right hand.
- Trap the aggressor's right hand by grabbing his wrist or hand with your right hand and applying pressure against your body and the pistol to keep it in its holster.
- Step back with your right foot and pivot sharply to your right, off balancing the aggressor, so that you are next to him. Bring your left arm perpendicular to, and down on, the aggressor's elbow.
- At the same time, straighten the aggressor's arm and apply an armbar. The aggressor's trapped arm should be straight across your torso. From here you should be able to control and take the aggressor down. In the case that you are unable to control the situation in this manner and the aggressor is fighting to straighten up, you should execute the following steps:
  - Execute a crossface, by turning his head to the opposite direction and off balancing him.
  - Grab the aggressor's face, apply both back and down pressure to his face, and step back with the left foot in order to take him to the ground. Maintain control of the aggressor's right hand at your right side the entire time.
- If you are unable to crossface your aggressor, the following steps are used:
  - Grab the aggressor's right shoulder with your left hand.
  - Digging your fingers into his brachial plexus tie-in, bring him up far enough to crossface your aggressor.
  - Follow-on by releasing the aggressor as he falls to the ground and transition to your firearm.

See figure 5-19 on 5-26.

Wristlock Technique

The wristlock technique is used when an aggressor grabs your pistol in the holster with his right hand.

Technique
- Begin with the aggressor facing you and grabbing the pistol in your holster with his right hand.
- With your right hand, trap the aggressor's right hand by grasping his hand and apply pressure against your body and on the pistol to trap it in its holster.
- Step back with your right foot and pivot sharply to your right, off balancing the aggressor, so that you are next to him while bringing your left arm perpendicular to and down on his elbow.
- Maintain pressure on the aggressor's right elbow with your left elbow and maintain a slight bend at the waist. Pivot your left hand to trap the aggressor's right hand so that your forearm is parallel with the aggressor's attacking arm.

Figure 5-19. Armbar Technique.

~ Execute a wristlock using the following technique.

•Place your left thumb on the back of the aggressor's right hand so that your knuckles are facing to your left.

•With your left hand, hook your fingers across the fleshy part of the aggressor's palm.

•Incorporate your second hand into the wristlock, exert downward pressure with your thumbs, and rotate the aggressor's hand to your left. Step back with your left foot, pivoting to your left to off balance the aggressor and drive him to the ground.

Marine Corps Martial Arts Program 5-27

~ Follow-on by releasing the aggressor as he falls to the ground and transition to your firearm. See figure 5-20.

Figure 5-20. Wristlock Technique.

Same Side Grab: From Front

The same side grab from the front technique can be used when the aggressor grabs your pistol in the holster with his left hand.

Chapter 5: Brown Belt · For Official Use Only

Technique

- Begin with the aggressor facing you and grabbing the pistol in your holster with his left hand.
- With your right hand, trap the aggressor's right hand by grasping his hand. Apply pressure against your body and on the pistol to trap it in its holster.
- Step back with the right foot, rotating the hip (right side) to the rear, at the same time with your left hand, strike the aggressor in the upper torso area (simulating a strike to the trachea). Striking surface will be the webbing between thumb and pointer finger.
- Follow up by releasing the aggressor's hand, doubling the distance and transition to your firearm.

See figure 5-21.

Figure 5-21. Same Side Grab: From Front.

Same Side Grab: From Rear

The same side grab from the rear technique can be used when an aggressor grabs your pistol in the holster with his right hand.

Technique
- Begin with the aggressor behind you and grabbing the pistol in your holster with his right hand.
- With your right hand, trap the aggressor's right hand by grasping his hand. Apply pressure against your body and on the pistol to trap it in its holster.
- Execute a reverse wristlock using the following technique:
    - Place the palm of your right hand on the back of the aggressor's right hand and wrap your fingers across the fleshy part of his palm below his little finger.
    - Twist the aggressor's hand to the right while placing the hand against your chest. Bring up the left hand to support the right hand by grabbing the aggressor's hand in between both hands, mimicing praying. Apply downward pressure on his hand against the chest. Leave the aggressor's hand on the chest to fully control him and to gain leverage. The aggressor's hand should be rotated 90-degrees so that his palm is facing left.
    - Step back with your right foot to maintain better balance and lean forward to use body weight to add additional pressure to the joint.
    - Follow up by releasing the aggressor and transition to your firearm.

See figure 5-22 on page 5-30.

Figure 5-22. Same Side Grab: From Rear.

Section VII
Firearm Disarmament

The purpose of firearm disarmament is to disarm the aggressor and control the situation by performing counter to the pistol, either front or rear.

Refer to appendix A for corresponding safeties 1, 2, 6, 8, 9, and 13.

CAUTION
To prevent injury during training, the aggressor should grip below the trigger housing only, keeping his finger out of the trigger housing at all times.

Counter to Pistol: Front

The counter to the pistol to the front technique is performed when you are unarmed and your aggressor is in front of you pointing a pistol at your head or chest. The technique is the same if the aggressor sticks the pistol under your chin.

Technique
- Begin with the aggressor presenting a pistol in his right hand to your chest. The pistol must be touching or very close to you for this technique to work.
- Place your hands in a submissive posture even with your shoulders, elbows into the body, and palms facing the aggressor. Make a submissive verbal statement.
- Clear your body from the line of fire by rotating your torso bringing the right shoulder back and, at the same time, grabbing the aggressor's wrist in a C-grip with your left hand pushing the weapon offline. Maintain control of the aggressor's arm.
- Step into the aggressor with your right foot and grab the weapon with your right hand in a C-grip by placing your thumb underneath the pistol and your fingers over top of the pistol. This rotates the pistol in the aggressor's hand. An incidental forward horizontal elbow strike to the aggressor is possible while removing the pistol from his grip.
- Step back to create distance from the aggressor and transition to employ follow-on actions by performing an expedient press check on the firearm.

See figure 5-23 on page 5-32.

Counter to Pistol: Rear

The counter to the pistol to the rear technique is performed when you are unarmed and your aggressor is behind you pointing a pistol at the back of your head or your back.

Technique
- Begin with the aggressor presenting a pistol in his right hand to your back. The pistol must be touching or very close to you for this technique to work.

Figure 5-23. Counter to Pistol: Front.

- ~ Place your hands in a submissive posture even with your shoulders, elbows into the body, and palms facing away from you. Make a submissive verbal statement and take a quick look in order to identify which hand the weapon is in.
- ~ Turn into the aggressor with your left foot, pivoting on your right foot while rotating your torso. The movement with the left foot should be deep enough to set up for the subsequent leg sweep. Use your left forearm to knock the weapon offline, doubling the distance between the weapon and your body, clearing you from the weapon's line of fire. Keep your hands up.
- ~ Quickly over hook the aggressor's right arm with your left arm trapping it in your armpit between the torso and the biceps, execute a right chin jab/palm heel strike.
- ~ Execute a leg sweep taking the aggressor to the ground. Maintain control of the aggressor's right arm.
- ~ Place your right hand on the aggressor's left shoulder or upper arm and your left palm on your right wrist to further control his arm and to affect an armbar.
- ~ Execute an armbar and continue to exert steady pressure against the arm to force the aggressor's release of the weapon. Use your right knee to control his hips and/or abdominal area.
- ~ Once the aggressor releases the weapon, release his arm, retrieve the weapon, and step back to create distance from him. Execute follow-on actions by performing an expedient press check in order to ensure that a round is chambered in the firearm.

See figure 5-24 on page 5-34.

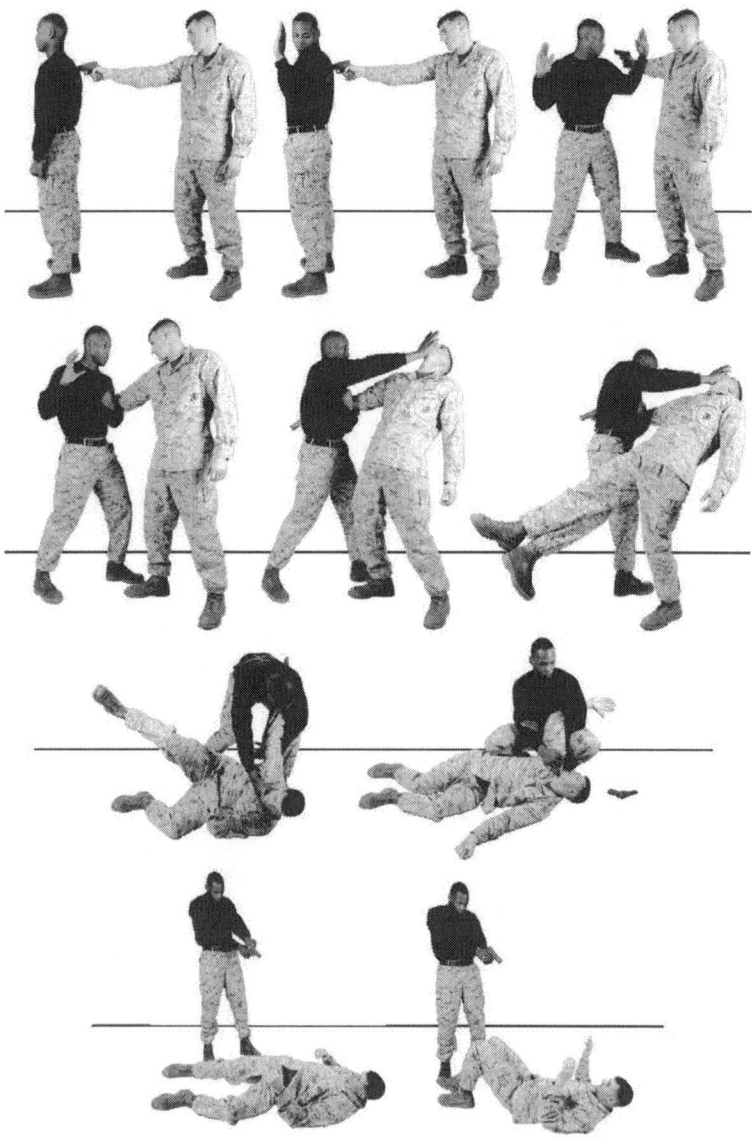

Figure 5-24. Counter to Pistol: Rear.

Section VIII
Knife Techniques

The purpose of knife fighting is to kill or cause enough damage and massive trauma to stop the aggressor by executing a block for a reverse strike or a block for a straight thrust.

Refer to appendix A for corresponding safeties 1, 2, 3, and 14.

Block for a Reverse Strike

The block for a reverse strike is effective against an aggressor that is excuting a reverse strike.

Technique
- From the modified basic warrior stance, move forward right, outside the arc of the aggressor's attack.
- Block the attack with your left arm and knife. Keep your left arm bent so that your forearm makes contact with the aggressor's triceps. Make contact with the knife and slash the aggressor's forearm.
- Maintain control of the aggressor's arm and follow through with at least three follow-on techniques to the available target areas.

See figure 5-25 on page 5-36.

Figure 5-25. Block for a Reverse Strike.

Block for a Straight Thrust

The block for a straight thrust is effective against an aggessor that is executing a straight thrust.

▃▃ Technique
- From the modified basic warrior stance, bend at the waist, moving your hips backwards and jumping backwards with both feet moving away from the attack, hollowing out. Thrust both hands out forcefully, with your left hand making contact palm side down on the aggressor's forearm. Your right hand holds the knife, making contact on the aggressor's forearm with the knife. The knife is parallel to the fingers of your left hand.
- Slash through the aggressor's right arm, maintain control of it with your left hand and follow through with at least three follow-on techniques to the available target areas.

See figure 5-26.

Figure 5-26. Block for a Straight Thrust.

This Page Intentionally Left Blank.

CHAPTER 6

Black Belt

The Black Belt is the fifth belt ranking within MCMAP. Upon qualifying as a Brown Belt, all Marines are highly encouraged to continue MCMAP sustainment training in order to advance to Black Belt. At an advanced level, the purpose and principles remain the same as outlined in all of the previous belts.

Black Belt Requirements.

Prerequisites	Recommendation of reporting senior
	Complete Brown Belt sustainment and integration training
	Appropriate level PME complete
Training Hours	Minimum of 20.7 hours, excluding remedial practice time and testing
Sustainment Hours	Minimum of 20 hours of sustainment, excluding integration training time and practice time for testing

Section I
Bayonet Techniques

The purpose of bayonet techniques is to disable or kill the aggressor.

Refer to appendix A for corresponding safeties 1, 2, 4, 7, 8, 9, and 14.

Bayonet Training: Stage One

Stage one of bayonet training focuses on basic posture, movement, and sequence of movements against a compliant target:

- The placement of the right hand on the pistol grip allows greater generation of force when executing the forward thrust, which is the primary offensive bayonet technique. Additionally, it allows you to transition immediately to assault fire as needed by moving the finger back to the trigger. For safety reasons, the finger is kept off of the trigger when executing bayonet techniques, this prevents an accidental discharge and protects the finger.
- The left hand is placed on the handguards in a position that is comfortable for the individual. If the hand is placed too far forward it causes an over extension of the left hand and mitigates some of the power and control that is generated with the offensive bayonet techniques.
- The buttstock locked into the hip is critical because it provides stability during a bayonet engagement when locking up with an aggressor or ensuring optimum power is generated when executing the thrust or any of the other offensive bayonet techniques.
- The blade of the bayonet is always pointed at the aggressor in order to facilitate a rapid engagement. Movement should be within an imaginary box that is shoulder-width wide from your neck to your waist.

Bayonet Training: Stage Two

Stage two of bayonet training adds the movement against multiple aggressors and integrates multiple weapons systems and bayonet techniques.

Bayonet Training: Stage Three

Stage three of bayonet training develops the ability to react effectively in the dark. Using your eyes effectively at night requires the application of the principles of night vision such as dark adaptation, off center vision, and scanning. Applying night vision principles alone will not guarantee a victory in a low light environment. It is necessary to combine these techniques with all of the others that you have practiced such as movement, posture, and technique. Low light engagements also require you to adjust your approach and close speeds due to uncertain terrain.

Dark Adaptation

Dark adaptation allows the eyes to become accustomed to low levels of illumination. It takes approximately 30 minutes for you to be able to distinguish objects in dim light.

Off Center Vision

Off center vision is the technique that allows your attention to be focused on an object without looking directly at it. When you look directly at an object, the image is formed on the cone region of your eye, which is not sensitive at night. When you look slightly off center (optimum is usually 6 to 10 degrees of an object), the image is formed on the area of your retina containing rod cells, which are sensitive in darkness.

Scanning

Scanning uses off center vision to observe an area or an object. Since rod cells only retain an image for 4 to 10 seconds, you must shift your eyes slightly so fresh rod cells are used. This is accomplished by moving your eyes in short, abrupt, irregular movements over and around your primary target.

Common Error

It is common for the Marine to wait too long and thrust too late, also known as cocking the weapon. When at closing distance, you should begin to thrust the bayonet trainer so that you can engage the aggressor at the longest possible distance. As a result of waiting too long to thrust, you get caught up and entangled with your aggressor, the bayonet trainer, or both and tend to pull the weapon back too far with your arms. The results put the Marine at a serious disadvantage for the following reasons:

- It completely disrupts the Marine's momentum. Often the Marine will find it necessary to completely stop and reverse direction of movement to clear the blade.
- Pulling the weapon back too far puts it in the Marine's weaker range of motion, which negatively impacts the power and ability to execute follow-on bayonet techniques.
- Cocking the weapon at any time serves to telegraph the Marine's intent to thrust the bayonet.

Allowing the aggressor to turn this into two, one-on-one engagements increases the chances of the Marines being separated and killed without the support or cover of his fellow Marine. While sometimes unavoidable, this is the least desirable of all outcomes.

Section II
Sweeping Hip Throw

A sweeping hip throw is particularly effective if the aggressor is moving forward or pushing on you. The sweeping hip throw is used to take your aggressor to the ground if your aggressor widens his stance in an attempt to prevent you from executing the hip throw. Execution of the sweeping hip throw uses the aggressor's forward momentum. This is accomplished by sweeping your aggressor's supporting leg and simultaneously continuing to take him to the ground while you remain standing. When teaching the sweeping hip throw, walk the students through the technique, step by step, working on the proper body position and execution.

Refer to appendix A for corresponding safeties 1, 2, 6, 13, and 15.

Technique
~ Stand facing the aggressor in the basic warrior stance.
~ Grab the aggressor's right wrist with your left hand.
~ Step forward with your right foot even with or slightly inside of the aggressor's right foot. Your heel should be between his feet, and your toes should be even with the aggressor's toes.
~ Step back with your left foot, rotating on the ball of your right foot. Your feet should be in between the aggressor's with your knees bent.
~ At the same time, rotate your waist and hook your right arm around the back of the aggressor's body anywhere from his waist to his head, depending on the size of the aggressor. If the aggressor is shorter than you, it may be easier to hook your arm around his head.
  • Hand placement should allow you to control the aggressor and pull him in close to you.
  • Your back side and hip should be up against the aggressor.
~ Rotate your hip up against the aggressor. Your hips must be lower than his. Use your right hand to pull the aggressor up on your hip to maximize contact.
~ Pull the aggressor's arm across your body and, at the same time, slightly lift him off of the ground by bending at the waist, straightening your legs.
~ Once the aggressor starts to come off of the deck, forcefully sweep his upper right thigh back with your right leg.
~ At the same time, continue to pull the aggressor's right arm forcefully to the left across your body to assist in bringing him to the deck.
  • If the aggressor cannot be easily lifted, your body positioning is not correct.
  • Students will execute a minimum of 10 fit ins per throw.
See figure 6-1.

Marine Corps Martial Arts Program

Figure 6-1. Sweeping Hip Throw.

Section III
Ground Fighting

The purpose of ground fighting is to apply the techniques that will allow you to get back to your feet as quickly as possible and cause damage to the aggressor by executing the face rip from the guard, the straight armbar from the scarf hold, and the bent armbar from the scarf hold.

Refer to appendix A for corresponding safeties 1, 2, 6, and 10.

Face Rip From the Guard

The face rip from the guard technique damages your aggressor and assists you in transitioning back to your feet as quickly as possible gaining a tactical advantage. It is executed when you are in the aggressor's guard and the aggressor is trying to damage you by striking at your face and head.

Technique

- Lay on your back with the aggressor kneeling in your guard position.
- The aggressor is trying to cause damage to you by striking your head.
- Pull down on the back of the aggressor's neck or head with both hands so that his head is on or next to your right shoulder, and his chin is facing outboard. Hug the back of the aggressor's neck to keep him on your chest and to ensure that you have control of his head.
- Reach around the back of the aggressor's neck with your left arm and grab his chin with your left hand.
- Place your right hand on the right side of the aggressor's chin.
- Twist the aggressor's neck by pulling to the left with your left hand and pushing up with your right hand.
- Push off of the ground with your right foot while blocking the aggressor's right leg with your left leg. Continue to exert pressure on his neck, coming to the mount position over top of him.
- Continue to apply pressure to the aggressor's chin and face with your right hand and follow-on with strikes with your left hand.

See figure 6-2.

Figure 6-2. Face Rip From the Guard.

Straight Armbar From a Scarf Hold

A straight armbar from a scarf hold technique causes pain compliance in your aggressor and assists you in transitioning back to your feet as quickly as possible in order to gain a tactical advantage.

Technique

~ Begin with the aggressor lying on his back. You sit to his right with your back/right side against the right side of his chest/ribs. Place the majority of your weight on your right hip. Wrap your right arm around the back of the aggressor's neck and grasp his right triceps with your left hand.

~ Spread your legs to better maintain your balance and to reduce the chance of the aggressor rolling you.

~ Release the aggressor's triceps with your left hand and grab his right wrist. Apply downward pressure with your left hand on his arm so that it is straight across your right leg. Maintain control of his head with your right arm.

~ Drape your left leg over the aggressor's right forearm and apply downward pressure by forcing your left knee toward the deck, simultaneously applying upward pressure with your right leg in a scissoring motion. Maintain control of your aggressor's arm.

~ Keep your head and chin tucked to avoid being grabbed or choked by the aggressor's free hand. During training, apply slow, steady pressure giving your training partner a chance to tap out. For joint destruction, quickly scissor legs while maintaining control of your aggressor.

See figure 6-3.

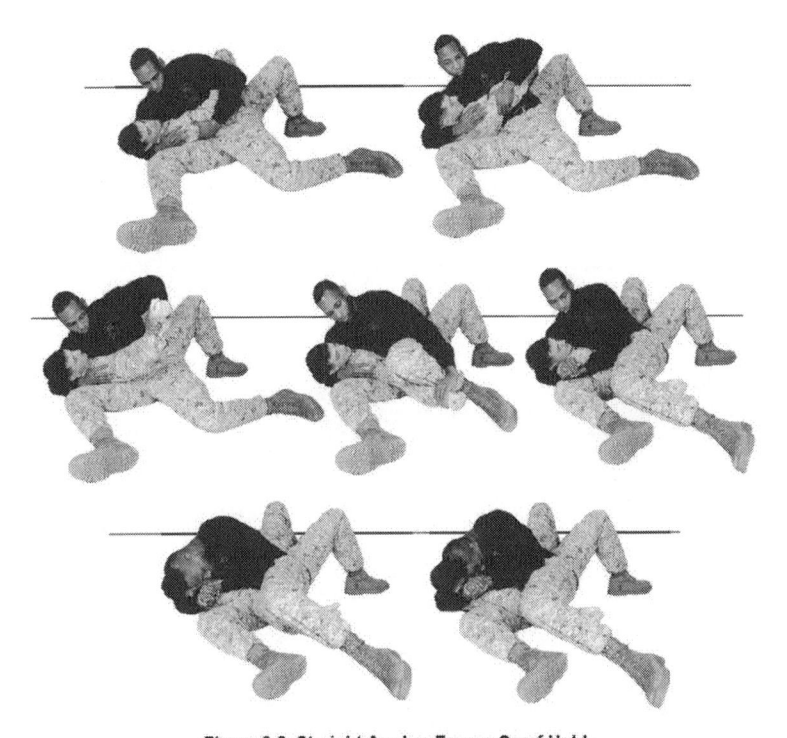

Figure 6-3. Straight Armbar From a Scarf Hold.

Bent Armbar From a Scarf Hold

A bent armbar from a scarf hold technique causes pain compliance in your aggressor and assists you in transitioning back to your feet as quickly as possible in order to gain a tactical advantage in a fight. Attempt to put the aggressor in a straight armbar of the straight armbar from a scarf hold technique as shown in figure 6-3. The aggressor will often try to bend his arm to avoid the straight armbar from a scarf hold technique.

Technique

~ Elevate your right knee over the aggressor's right wrist and trap his wrist in the bend of your right knee.

~ Press your right knee back to the ground while drawing your right foot toward your buttocks. At the same time, clasp your hands together and pull up on the aggressor's head to apply additional pressure to the shoulder.

~ Keep your head and chin tucked to avoid being grabbed or choked by the aggressor's free hand. During training, apply slow, steady pressure giving your training partner a chance to tap out. For joint destruction, quickly draw your legs back and jerk up on your aggressor's head.

See figure 6-4.

Figure 6-4. Bent Armbar From a Scarf Hold.

Section IV
Unarmed Manipulation: Neck-Crank Takedown

Marines operate within a continuum of force, particularly in support of peacekeeping- or humanitarian-type operations. In these situations, Marines must act responsibly to handle situations without resorting to deadly force. Unarmed restraints and manipulation techniques, such as a neck-crank takedown, can be used to control an aggressor and gain the tactical advantage.

Refer to appendix A for corresponding safeties are 1, 2, 6, and 10.

Technique
- Begin in a static position with your feet in line and shoulder-width apart in front of the aggressor. Step in with the right foot while quickly placing your left hand behind the aggressor's head and firmly grasping his upper neck/lower head. At the same time, quickly place your cupped, right palm on the aggressor's chin with your fingers extended across the left side of his face so that he cannot pull away.
- Pull your left hand down and to the left as you forcefully push the aggressor's chin up and to the right to off balance him.
- Step back with your left foot and continue to apply pressure to the aggressor's neck, forcing him to the deck.
- Once the aggressor is on the deck, maintain control by applying pressure, keeping his head on the deck. Additionally place your knee on your aggressor's shoulder for control.
- Return to the basic warrior stance, creating a safe distance from the aggressor and maintaining awareness of your surroundings.

See figure 6-5 on page 6-12.

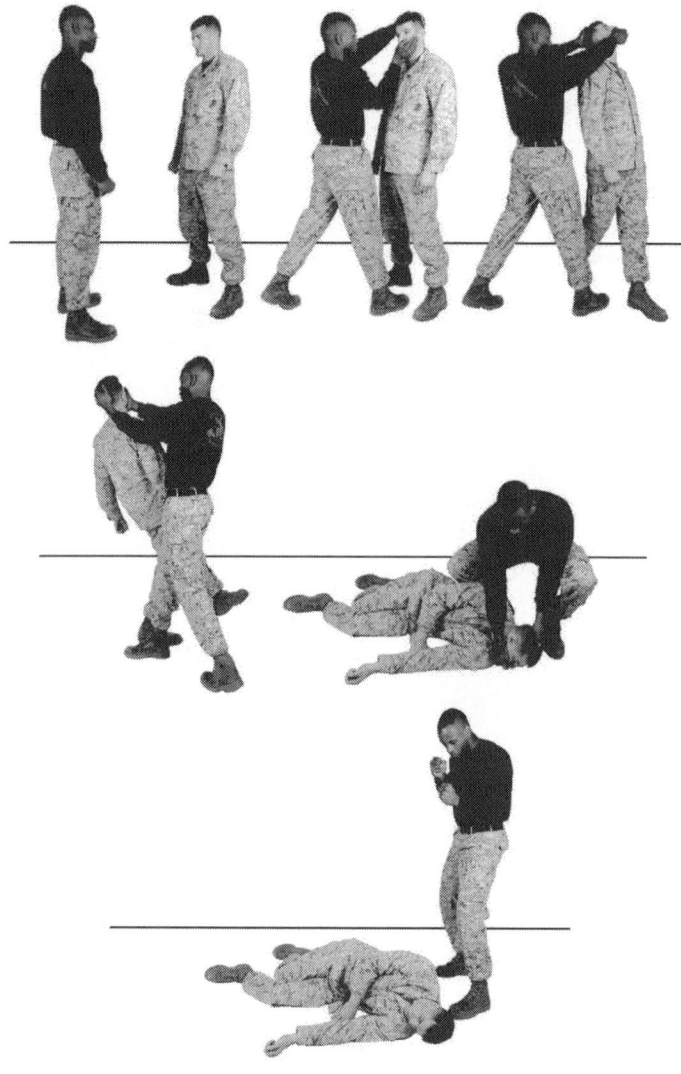

Figure 6-5. Neck-Crank Takedown.

Section V
Chokes

The purpose of a choke is to render your aggressor unconscious or gain control of a close combat situation through less than lethal force. Chokes are performed by either closing off the airway to the lungs, thereby preventing oxygen from reaching the heart or by cutting off the blood flow to the brain.

When executed properly, a blood choke takes between 8 to 13 seconds for the aggressor to lose consciousness. The air choke is least preferred because it takes longer to render the aggressor.

A blood choke, such as the triangle choke and the guillotine choke is performed on the carotid artery that carries oxygen-enriched blood from the heart to the brain. The carotid artery is located on both sides of the neck. When executed properly, a blood choke takes between 8 and 13 seconds for the aggressor to lose consciousness. The blood choke is the preferred choke because the intended effect is for the aggressor to quickly lose consciousness, ending the fight. A blood choke is used to render your aggressor unconscious or gain control of a close combat situation through less than lethal force.

An air choke is performed on the windpipe or trachea, cutting off the air to the lungs and heart. When executed properly, an air choke takes between 2 and 3 minutes for the aggressor to lose consciousness. Due to the length of time it takes to stop the fight with an air choke, air chokes are not recommended and will not be taught as part of MCMAP. This section covers the triangle choke and the guillotine choke.

Refer to appendix A for corresponding safeties 1, 2, 6, 10, 11, and 12.

Triangle Choke

The purpose of the triangle choke is to render an aggressor unconscious and quickly get back to your feet to gain the tactical advantage. This technique is executed if the aggressor is on top of you and your legs are wrapped around the aggressor's waist. In the following scenario, the aggressor is trying to pass your guard position and gain the tactical advantage.

Technique
- Begin by lying on your back with the aggressor kneeling between your legs in your guard position.
- The aggressor tries to pass your guard position by reaching back with his right arm and attempting to throw your left leg over his head.
- As the aggressor tries to throw your left leg over his head, turn your body slightly to the left, quickly place the back of your left knee along the right side of the aggressor's neck.
- Bend your left knee so that your calf is applying pressure on the back of your aggressor's neck. If necessary, elevate your hips slightly to help achieve the position.
- With both hands, grasp your aggressor's left wrist and pull it forcefully toward your left hip.
- Maintain control of your aggressor's left arm, lift your right leg off of the deck, and place the back of your right knee on the top of your left ankle.

~ Exert pressure on your aggressor's neck by pushing your right heel toward your buttocks. You can make this technique more effective by pulling on the back of your aggressor's head with both hands and thrusting your hips up.

See figure 6-6.

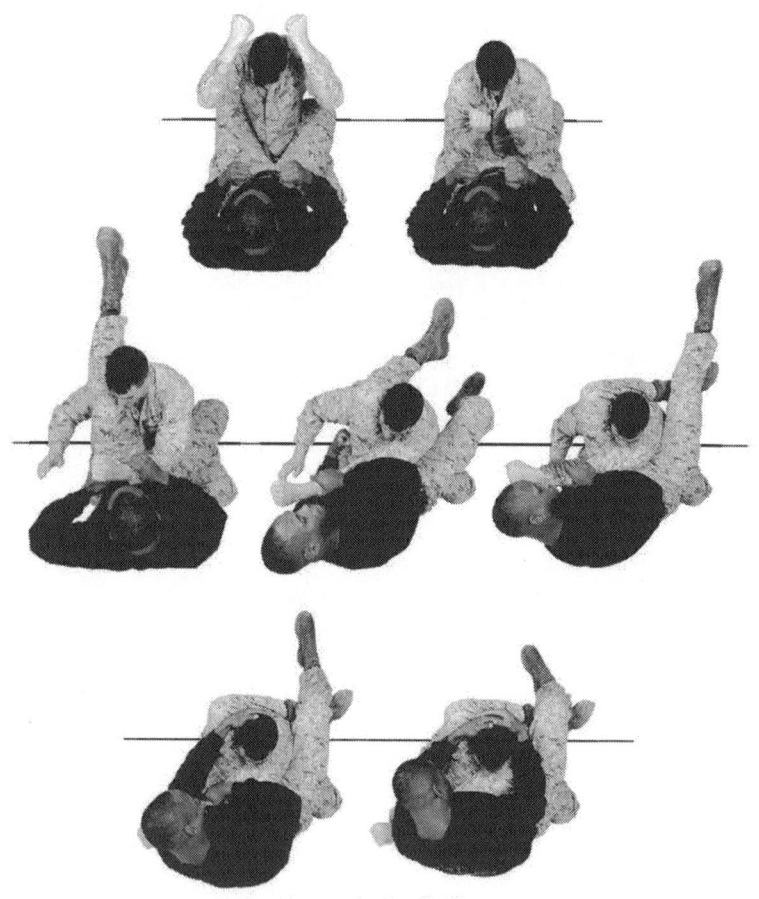

Figure 6-6. Triangle Choke.

Guillotine Choke

The purpose of a guillotine choke is to render an aggressor unconscious and quickly get back on your feet to gain the tactical advantage. This technique is performed when the aggressor is trying to tackle you by either grabbing both of your legs or grabbing around your waist and forcing you to the ground to gain a tactical advantage. This technique can be performed from either the standing position or from the ground.

Technique
- Begin by standing facing each other.
- As your aggressor tries to tackle you, wrap your right arm around his neck and clasp your hands together, then wrap your right leg around his left leg.
- By arching your back straight and pulling upward with the clasp of your hands, the force against the neck will cause your aggressor to choke out.
- If the engagement reaches the ground, wrap both legs around your aggressor's body (guard position), maintaining a tight clasp around the aggressor's neck.
- Use your ankles to separate the aggressor's legs. Make sure that you arch your back and apply pressure with your legs and arms at the same time, causing the aggressor to choke out.

See figure 6-7.

Figure 6-7. Guillotine Choke.

Section VI
Knee Bar

The purpose of the knee bar is to apply joint manipulation to the aggressor's knee in order to escape from a hold by executing the rolling knee bar or the sitting knee bar to gain a tactical advantage.

Refer to appendix A for corresponding safeties 1, 2, 5, 6, 10, and 13.

Rolling Knee Bar

The purpose of rolling the knee bar is to apply joint manipulation to the aggressor's knee in order to escape from a hold and gain a tactical advantage. This technique requires that your arms remain free.

Technique

- The aggressor executes a rear bear hug leaving your arms free.
- Execute a rear horizontal elbow strike to loosen the aggressor's grasp.
  - Horizontal elbow strikes are aimed at the head, but for training purpose they will be thrown in the air, do not make contact with your training partner.
  - Other distracters such as the vertical stomp to instep and the rear head-butt can be executed in a live situation.
- Step out slightly with your right foot, bending at the knees and the waist.
- Reach between your legs with both hands and firmly grasp the aggressor's right leg at or below the knee.
- Execute a forward shoulder roll and pull the aggressor's right leg close against your torso.
- Upon completing the forward shoulder roll you should be lying on your hip with the aggressor's leg held tight against your torso.
  - Your left leg should be between the aggressor's legs, tight to his groin, your left foot supported by your right foot. Keep your knees bent and pinched together so you can control the aggressor's upper leg.
  - The aggressor's knee should be at or slightly above your groin.
- Maintain control of the aggressor's leg and arch your hips into his knee while pulling back with your whole body. Your arms, which are holding his leg to your torso, and your feet will also cinch the aggressor's leg and pull.
- Thrust your hips forward quickly while yanking back on the aggressor's leg for joint destruction. During training, apply slow, steady pressure for the submission with your training partner.

See figure 6-8.

Marine Corps Martial Arts Program 6-17

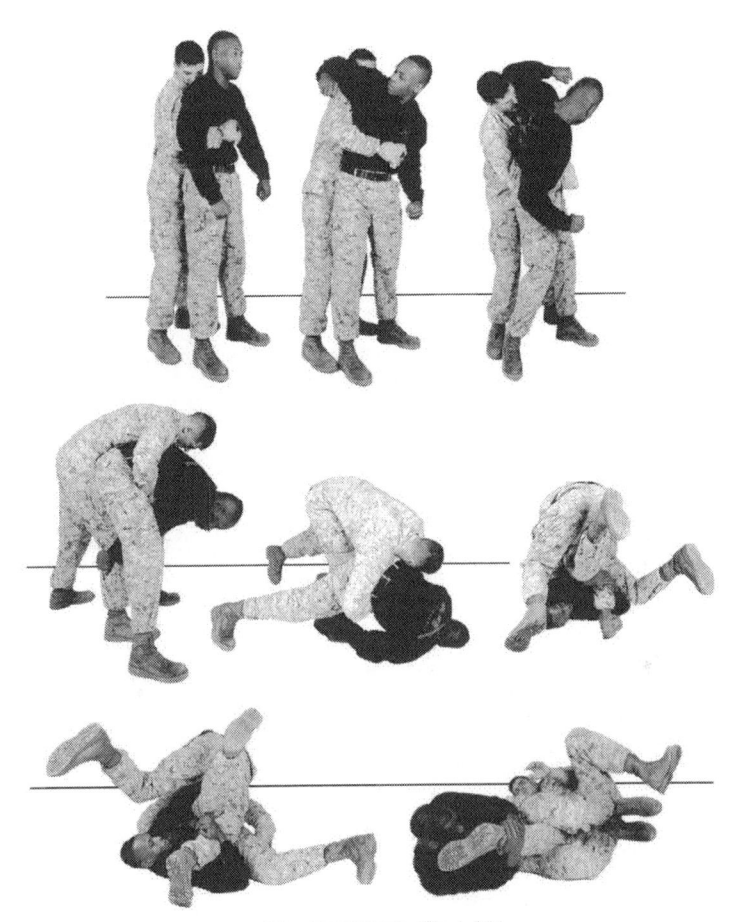

Figure 6-8. Rolling Knee-Bar.

Chapter 6: Black Belt

Sitting Knee Bar

The purpose of the sitting knee bar is to apply joint manipulation to the aggressor's knee to escape from a hold and gain a tactical advantage. This technique requires that your arms remain free.

Technique

- The aggressor executes a rear bear hug leaving your arms free.
- Execute a rear horizontal elbow strike to loosen the aggressor's grasp.
  - Horizontal elbow strikes are aimed at the head, but for training purpose they will be thrown in the air, do not make contact with your training partner.
  - Other distracters such as the vertical stomp to instep and the rear head-butt can be executed in a live situation.
- Step out slightly with your right foot, bending at the knees and waist.
- Reach between your legs with both hands and firmly grasp the aggressor's right leg at or below the knee.
- Sit back on your aggressor's leg while simultaneously pulling the aggressor's leg close against your torso.
- After completing the sitting motion, you should be laying on your hip with the aggressor's leg held tight against your torso.
  - Your left leg should be between the aggressor's legs, tight to his groin, left foot supported by your right foot. Keep your knees bent and pinched together so you can control the aggressor's upper leg.
  - The aggressor's knee should be at or slightly above your groin.
- Maintain control of the aggressor's leg and arch your hips into his knee while pulling back with your whole body. Your arms, which are holding the leg to your torso, and your feet will also cinch the aggressor's leg and pull.
- Thrust your hips forward quickly while yanking back on the aggressor's leg for joint destruction. During training, apply slow, steady pressure for the submission with your training partner.

See figure 6-9.

Marine Corps Martial Arts Program 6-19

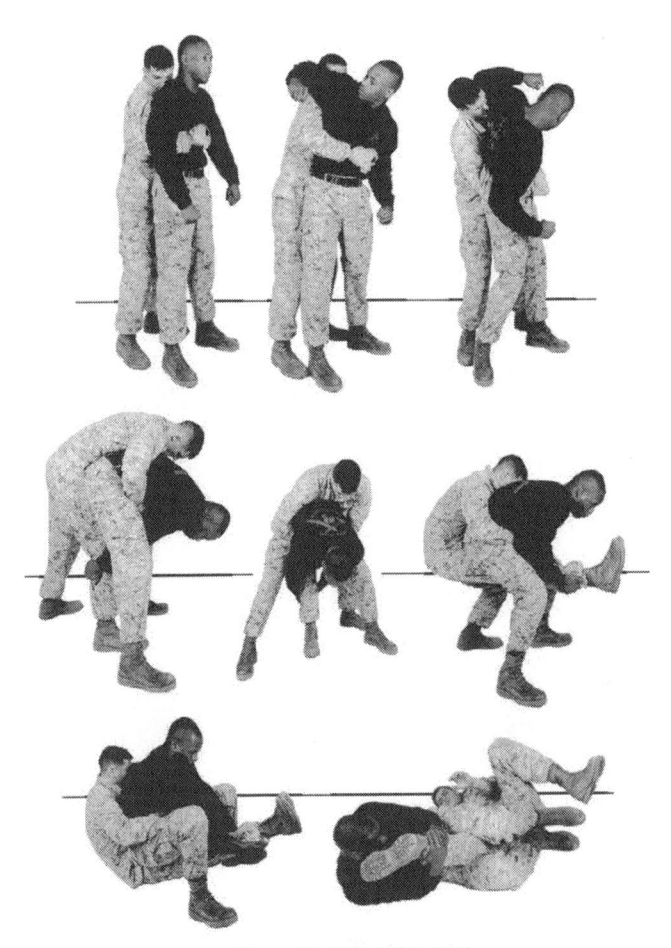

Figure 6-9. Sitting Knee Bar.

Section VII
Counter to Pistol to the Head

Firearm disarmament techniques are designed to counter a confrontation with a pistol while permitting you to gain the tactical advantage against an aggressor.

Refer to appendix A for corresponding safeties 1, 2, 6, 8, 9, 10, and 13.

Counter to Pistol to the Head: Two Handed

This technique is performed when you are unarmed and your aggressor has a pistol pointing at your head. The pistol must be in close distance to you for this technique to be effective.

Technique

- Begin at close range with the aggressor pointing a pistol at your head. The aggressor is holding the pistol with both hands.
- Assume a submissive posture, your hands about chest high and palms out. Place your hands as close to the weapon as possible without raising the aggressor's suspicion. Make a submissive verbal statement.
- Grab your aggressor's wrists with both hands and thrust them upward. At the same time, drop your body and head down quickly, while bending your knees. You must do this at the same time to double your distance from the barrel of the pistol and to clear your head out of the line of fire.
  - Maintain control of the aggressor's arm(s).
  - It does not matter which hand the aggressor is holding the weapon with, always grab his wrist(s) and clear your head in the same manner.
- Step your right foot to the outside of the aggressor's right foot and trap his arm in your right armpit. Rotate your torso and thrust your hip into him to off balance him.
- Pull your right elbow in tight to your side to trap the aggressor's arms. Retain a firm grip on the aggressor's wrist(s) and do not allow them to rotate in your grasp.
- Maintain your right over hook; grab the pistol with your left hand, rotating it out of the aggressor's hand. Make sure the muzzle is never pointed at you and that you keep your hand clear of it.
- Use your hips and your upper body leverage to off balance the aggressor and to put space between you and the aggressor. Turn toward the aggressor, perform an expedient press check, and prepare for follow-on actions.

See figure 6-10.

Figure 6-10. Counter to Pistol to the Head: Two Handed.

Counter to Pistol to the Head: One Handed

This technique is performed when you are unarmed and your aggressor has a pistol pointing at your head. The pistol must be in close distance to you for this technique to be effective.

Technique
- Begin at close range with the aggressor pointing a pistol at your head. The aggressor is holding his pistol with his right hand and his left arm is down.
- Assume a submissive posture, your hands about chest high and palms out. Place your hands as close to the weapon as possible without raising the aggressor's suspicion. Make a submissive verbal statement.
- Grab your aggressor's wrist with both hands and thrust upward. At the same time, drop your body and head down quickly, while bending your knees. You must do this at the same time to double your distance from the barrel of the pistol and to clear your head out of the line of fire.
  - Maintain control of the aggressor's arm.
  - It does not matter which hand the aggressor is holding the weapon with, always grab his wrist(s) and clear your head in the same manner.
- Step with your right foot to the outside of the aggressor's right foot and trap his arm in your right armpit. Rotate your torso and thrust your hip into the aggressor to off balance him.
- Pull your right elbow in tight to your side to trap the aggressor's arm. Retain a firm grip on the aggressor's wrist and do not allow him to rotate in your grasp.
- Maintain your right over hook; grab the pistol with your left hand, rotating it out of the aggressor's hand. Make sure that the muzzle is never pointed at you and that you keep your hand clear of it.
- Use your hips and your upper body leverage to off balance the aggressor and to put space between you and the aggressor. Turn toward the aggressor, perform an expedient press check, and prepare for follow-on actions.

See figure 6-11.

Marine Corps Martial Arts Program 6-23

Figure 6-11. Counter to Pistol to the Head: One Handed.

Section VIII
Upper Body Strikes

The purpose of an upper body strike is to stun the aggressor by using the cupped hand strike and the face smash, which sets the aggressor up for follow up techniques.

Refer to appendix A for corresponding safeties 1, 2, 3, and 4.

Cupped Hand Strike

Striking with the cupped hand concentrates power in a small part of the hand which, when transferred to the primary target, can have a devastating effect.

Striking Surface

The striking surface of the cupped hand strike is primarily the palm of the hand.

Target Areas of the Body

The primary target areas of the body are the neck, the face, the head, the ribs, the groin, and the kidneys.

Technique
- From the basic warrior stance, open your right hand about halfway, keeping your fingers and thumb together.
- Retract your right arm so that your hand is next to the right side of your face and neck. Your arm is bent at approximately a 45- to 90-degree angle. At the same time, rotate your right hip and right shoulder backwards.
- Thrust your hand forward while rotating your right hip and shoulder forward.
- Rotate your shoulder so that the concave portion of the cupped hand makes contact on the aggressor.
- Follow through the primary target area with your hand and return to the basic warrior stance.

See figure 6-12.

Face Smash

Striking with the face smash concentrates power in a small part of the hand which, when transferred to the primary target, can have a devastating effect.

Striking Surface

The striking surface is primarily the palm of the hand, and secondarily, the fingertips.

Target Areas of the Body

The primary target area of a face smash is the face.

Marine Corps Martial Arts Program

Figure 6-12. Cupped Hand Strike.

Technique

- From the basic warrior stance, open your right hand. Spread and slightly bend your fingers with muscular tension. The hand looks like it could be holding a grapefruit at this stage.
- Retract your right arm so that your hand is next to the right side of your face and neck. Your arm is bent at approximately a 45- to 90-degree angle. At the same time, rotate your right hip and right shoulder backwards.
- Thrust your hand forward while rotating your right hip and shoulder forward and forcefully step forward while pushing off on the ball of your right foot. Immediately bring your right foot up and return to the basic warrior stance.
- Contact should be made with the palm of the hand and the fingertips.
- Follow through the primary target area with your hand and rake the face of the aggressor with your fingers.
- When thrown vertically, the face smash comes straight down in an arcing motion. The face smash gets its power by moving your body in a linear line and taking a forceful step forward with the left foot, pushing off on the ball of the right foot, while rotating your hips and shoulder into the attack.

See figure 6-13.

Figure 6-13. Face Smash.

Section IX
Knife Techniques

The purpose of knife fighting is to kill or cause enough damage and massive trauma to stop an aggressor.

Refer to appendix A for corresponding safeties 1, 2, 3, and 14.

Lead Hand Knife

The purpose of a lead hand knife is to give you the ability to use more speed than power, it also gives you more space between you and the aggressor, especially if he is armed with an edged weapon. Your stance will be slightly different than your normal modified basic warrior stance.

Technique

- ~ The weak side hand serves as a vertical shield protecting the ribs, head, and neck. It will also be used to strike when bulldogging your aggressor.
- ~ Place your strong side leg forward, your strong side elbow slightly bent with the blade and tip pointing forward toward the aggressor's head. This position serves as an index point, where all lead hand techniques are initiated.
- ~ Your strong side forearm will be almost parallel to the deck and forward of your weak side elbow. The weapon will be held approximately chest high to belt level inside your box.
- ~ The weapon will be kept away from your body to facilitate quicker strikes and to control the distance.
- ~ Hold the knife in a hammer grip in your lead hand.

See figure 6-14.

**Figure 6-14.
Lead Hand Knife.**

Slashing Techniques

Slashing techniques distract the aggressor or cause enough damage so that you can close with him and apply more damaging techniques. Primary target areas are usually the limbs or any portion of the body that is presented. Black Belt vertical, forward, and reverse slashing techniques are identical to those in earlier belt levels, except these slashing techniques are performed with the strong side forward.

Vertical Slash

Technique
- Stand in the modified basic warrior stance facing the aggressor.
- Thrust your strong side hand out, and bring the weapon's edge straight down on the aggressor.
- Cut down through the aggressor's body.
- Return to the modified basic warrior stance.

See figure 6-15.

Figure 6-15. Vertical Slash.

The Forward Slash

Technique
- Stand in the modified basic warrior stance facing the aggressor.
- Extend your strong side hand to cut the aggressor with the blade.
- Rotate your palm up to make contact with the blade on the aggressor.
- Do not move your arm outside of the box (shoulder width across from your neck to your waistline).
- Upon contact, snap the wrist through the slashing motion to maximize contact with the blade on the aggressor.
- Continue cutting with the knife through the aggressor's body, from your strong side to weak side, in a forehand stroke.
- Maintain contact on the aggressor's body with the blade of the knife.

~ The movement ends with your strong side elbow or triceps against your body and the knife at your weak side, inside your box, and with the blade point oriented on the aggressor.
~ Return to the modified basic warrior stance.
See figure 6-16.

Figure 6-16. Forward Slash.

Reverse Slash

Technique
~ Stand in the modified basic warrior stance facing the aggressor.
~ Bend your strong side arm slightly, crossing your forearm to the weak side and forward of your body inside the box. Rotate your wrist palm down.
~ Extend your strong side hand to cut the aggressor with the knife blade.
~ Rotate your palm down to make contact with the blade on the aggressor.
~ Do not move your arm outside the box (shoulder width across from your neck to your waistline).
~ Upon contact, snap the wrist through the slashing motion to maximize contact with the blade on the aggressor.
~ Continue cutting with the knife through the aggressor's body, from your weak side to strong side, in a backhand stroke. Maintain contact on his body with the blade of the knife.
~ Return to the modified basic warrior stance.
See figure 6-17.

Thrusting Techniques

The primary objective when fighting with a knife is to insert the blade into an aggressor to cause massive damage and trauma. This is done with a thrusting technique. Thrusting techniques are more effective than slashing techniques because of the damage they can cause. Black Belt thrusting techniques are identical to those in earlier belt levels, except in these thrusting techniques are performed with the strong side forward.

Marine Corps Martial Arts Program

Figure 6-17. Reverse Slash.

Vertical Thrust

Technique
- Stand in the modified basic warrior stance facing the aggressor.
- Thrust your strong side hand toward the primary target, inserting the knife blade straight into the aggressor.
- Pull the knife out of the aggressor.
- Return to the modified basic warrior stance.

See figure 6-18.

Figure 6-18. Vertical Thrust.

Forward Thrust

Technique

- Stand in the modified basic warrior stance facing the aggressor.
- Thrust your strong side hand toward the primary target, palm down, inserting the knife blade into the aggressor.
- Once the knife is inserted, twist the blade by rotating your palm up. This enables the cutting edge of the blade to be in a position to further cut the aggressor in a follow up action.
- Turn the blade and cut your way out rather than pulling the knife straight out causes more damage and trauma to the aggressor.
- This action can be taken when thrusting to the aggressor's neck or abdomen region. However, if the aggressor is wearing body armor it may be difficult or impossible to bring the knife diagonally across his body.
- Drop your strong side elbow and bring the knife to the opposite side of the aggressor's body from where it was inserted. The movement ends with your strong side elbow or triceps against your body and the knife at your weak side, inside your box (shoulder width across from your neck to your waistline), and with the blade oriented toward the aggressor.
- At the same time, rotate your hips and shoulders downward to bring your body weight to bear on the attack.
- Return to the modified basic warrior stance.

See figure 6-19.

Figure 6-19. Forward Thrust.

Reverse Thrust

Technique

- Stand in the modified basic warrior stance facing the aggressor.
- Bend your strong side arm slightly, crossing your forearm to the weak side and forward of your body inside the box. Rotate your wrist palm side up.
- Thrust your strong side hand toward the primary target, palm side up, and insert the knife blade straight into the aggressor.

Marine Corps Martial Arts Program

- Once the knife is inserted, twist the blade by rotating your palm down. This enables the cutting edge of the blade to be in a position to further cut the aggressor in a follow up action.
- Turn the blade and cut your way out rather than pulling the knife straight out causes more damage and trauma to the aggressor.
- This action can be taken when thrusting to the aggressor's neck or abdomen region. However, if the aggressor is wearing body armor it may be difficult or impossible to bring the knife diagonally across his body.
- At the same time, rotate your hips and shoulders downward to bring your body weight to bear on the attack.
- Return to the modified basic warrior stance.

See figure 6-20.

Figure 6-20. Reverse Thrust.

Reverse-Grip Knife Techniques

Reverse-grip knife techniques are intended to give you different tactical options and different angles of attack. Your grip on the knife should be natural. The knife techniques described in the following subparagraphs are identical in purpose to those described earlier, with the exception of the changed grip. In reverse-grip knife techniques, the weak side is forward.

Grip

Your grip on the knife should be natural. Grasp the knife's grip with your fingers wrapped around the grip naturally, with the blade edge at a 90-degree angle, and the point oriented toward the deck. This is commonly known as a reverse hammer grip.

Stance

- The left hand serves as a vertical shield protecting the ribs or the head and neck.
- The right elbow is bent with the blade edge pointing forward toward the aggressor and the point toward the deck. This position serves as an index point, where all techniques are initiated.

- The weapon should be held at a level approximately from the top of the belt to chest high.
- The weapon should be kept in close to the body to facilitate weapon retention.

Reverse-Grip Forward Slash

Technique
- ~ Stand in the modified basic warrior stance facing the aggressor.
- ~ Extend your right hand in a hooking motion toward the aggressor, your forearm parallel with the deck, and your palm toward the deck to make contact on the aggressor with the knife blade.
- ~ Do not move your arm outside the box (shoulder width across from your neck to your waistline).
- ~ Upon contact, snap your wrist through the slashing motion to maximize contact with the blade on the aggressor.
- ~ Continue cutting with the knife through the aggressor's body, from your right to your left, in a forehand stroke.
- ~ Maintain contact on the aggressor's body with the blade of the knife.
- ~ The movement ends with your right elbow forward and forearm parallel to the deck and with the knife on the left side of your body, inside your box, and point oriented toward the aggressor.
- ~ Return to the modified basic warrior stance.

See figure 6-21.

Figure 6-21. Reverse-Grip Forward Slash.

Reverse-Grip Reverse Slash

Technique
- ~ Stand in the modified basic warrior stance facing the aggressor.
- ~ Start with your right elbow forward and forearm parallel to the deck and the knife on the left side of your body, inside your box, and point oriented toward the aggressor with your palm facing the ground.

- ~ Extend your right hand to make contact on the aggressor with the knife blade.
- ~ Rotate your palm up to make blade contact with the aggressor.
- ~ Do not move your arm outside the box (shoulder width across from your neck to your waistline).
- ~ Continue cutting with the knife through the aggressor's body, from your left to your right, in a backhand stroke. Maintain contact on the aggressor's body with the blade of the knife.
- ~ Return to the modified basic warrior stance.

See figure 6-22.

Figure 6-22. Reverse-Grip Reverse Slash.

Reverse-Grip Forward Thrust

Technique

- ~ Stand in the modified basic warrior stance facing the aggressor.
- ~ Thrust your right hand toward the primary target, palm side up, inserting the knife blade into the aggressor.
- ~ Once the knife is inserted, twist the blade by rotating your palm down. This enables the cutting edge of the blade to be in a position to further cut the aggressor in a follow up action.
- ~ Turn the blade and cut your way out rather than pulling the knife straight out causes more damage and trauma to the aggressor.
- ~ This action can be taken when thrusting to the aggressor's neck or abdomen region. However, if the aggressor is wearing body armor, it may be difficult or impossible to bring the knife diagonally across his body.
- ~ Raise your right elbow and bring the knife to the opposite side of the aggressor's body from where it was inserted. The movement ends with your right elbow forward and forearm parallel to the deck and the knife on the left side of your body, inside your box, and point oriented toward the aggressor.

- ~ At the same time, rotate your hips and shoulders downward to bring your body weight to bear on the attack.
- ~ Return to the modified basic warrior stance.

See figure 6-23.

Figure 6-23. Reverse-Grip Forward Thrust.

Reverse-Grip Reverse Thrust

Technique

- ~ Stand in the modified basic warrior stance facing the aggressor.
- ~ Start with your right elbow forward and forearm parallel to the deck and the knife on the left side of your body, inside your box, and point oriented toward the aggressor with your palm facing the deck.
- ~ Thrust your right hand toward the primary target, palm down, inserting the knife blade into the aggressor.
- ~ Once the knife is inserted, twist the blade by rotating your palm up slightly. This enables the cutting edge of the blade to be in a position to further cut the aggressor in a follow up action.
- ~ Turn the blade and cut your way out rather than pulling the knife straight out causes more damage and trauma to the aggressor.
- ~ This action can be taken when thrusting to the aggressor's neck or abdomen region. However, if the aggressor is wearing body armor, it may be difficult or impossible to bring the knife diagonally across his body.
- ~ At the same time, rotate your hips and shoulders downward to bring your body weight to bear on the attack.
- ~ Return to the modified basic warrior stance.

See figure 6-24.

Figure 6-24. Reverse-Grip Reverse Thrust.

Section X
Improvised Weapons

Improvised weapons, such as hard and soft garrotes, are made from locally available material. Unlike weapons of opportunity, improvised weapons are designed as a weapon for a specific function. The most common use of improvised weapons is for the silent removal of sentries during a raid, infiltration, reconnaissance mission, or other tactical scenarios as a means of incapacitating a potential prisoner or for use during a survival, escape, and evasion situation.

There are several methods for employing the various types of garrotes. The method chosen will depend on the tactical situation and intended results. The employment of a garrote is considered deadly force. In most cases, the garrote will be employed when silence and stealth are required. The approach and close with the aggressor should be deliberate and noiseless. Stay low with your body below the aggressor's line of sight. The entry should be rapid while applying explosive force to achieve the best results.

Refer to appendix A for corresponding safeties 1, 2, 6, 10, 11, 12, 13, and 14.

Garrote From the Rear

The garrote from the rear is intended to allow you to apply deadly force to an aggressor before he is aware of your presence.

With a Flexible Garrote
- Place your left hand palm side up in the center of your aggressor's back just below the neck.
- With your right hand, palm side down, loop the flexible garrote around your aggressor's neck from right to left, forming an X across your aggressor's back.
- With the inside of your right foot, collapse your aggressor down by striking the back of his right knee and riding it down to the ground.
- While staying close to your aggressor and still stepping on the back of his right calf, pull down and to the right with your right hand, while leaving your left hand in place until your aggressor is unconscious.
- See figure 6-25.

With a Hard Garrote
- Shoot the long end of a garrote across the aggressor's neck. From right to left, palm side up with the right ulna bone making contact against the aggressor's right carotid artery.
- With your left hand, grab the garrote palm side down placing the weapon across the aggressor's left carotid artery.
- While keeping your elbows in tight, squeeze with both arms in a vise motion until the aggressor is unconscious.
- See figure 6-26 on page 6-38.

Figure 6-25. With a Flexible Garrote.

Figure 6-26. With a Hard Garrote.

Garrote From the Front

The garrote from the front is intended to apply deadly force to an aggressor before he is aware of your presence. You may not be able to get fully behind an aggressor and, in some instances, when attempting to apply a garrote from the rear, the aggressor may become aware of your presence and turn to face you. In this situation you would use the garrote from the front.

With a Flexible Garrote
- With both arms straight out in front of you, move forward toward the aggressor's neck.
- Next, you will move around your aggressor's right side to his back while simultaneously wrapping your right arm around the aggressor's head so that the garrote ends up around the aggressor's neck forming an X on the aggressor's back. The right hand is palm side up, left hand is palm side down.
- Collapse the aggressor down to the ground by striking the back of the aggressor's right knee with the inside of your right foot.
- While staying close to your aggressor and still stepping on the back of his right calf, pull down and to the right with your right hand, while leaving your left hand in place until your aggressor is unconscious.
- See figure 6-27.

Figure 6-27. With a Flexible Garrote.

With a Hard Garrote

- Shoot the long end of the garrote across the back of the aggressor's neck, from your right to left palm side up with the right ulna across the aggressor's left carotid artery.
- With your left hand, grab the weapon palm side down, placing the weapon across the aggressor's right carotid artery.
- While keeping elbows in tight, squeeze with both arms in a vise motion until aggressor is unconscious.
- See figure 6-28.

Figure 6-28. With a Hard Garrote.

Appendix A

Training Safety Sheet

The following safeties are to be used for all belt levels.

I. Begin slowly and increase speed with proficiency.

II. Never execute techniques at full force or full speed.

III. When executing punches, ensure the joints are kept slightly bent to avoid hyperextension.

IV. Students will train under the supervision of a martial art instructor or on MAIT in accordance with all safety and logistical requirements and in the following stages:

- In the air (e.g., shadow boxing/imaginary targets).
  Note: Punches or strikes are not applied to the pads.
- On the striking pads or training tools.
- On the body during free sparring.

V. Prevent injuries during training; train break-falls in stages from the lowest position to the highest position.

VI. Techniques will be performed on a soft-footed area.

VII. Begin by executing bayonet techniques in the air. Ensure bayonets are sheathed. Ensure contact is not made with an aggressor when doing the target acquisition phase of air drills.

VIII. Before training with any firearms, unload and show clear will be conducted.

IX. When handling weapons, the following four safety rules apply:

- Treat every weapon as if were loaded.
- Never point your weapon at anything you do not intend to shoot.
- Keep your finger straight and off of the trigger until you intend to fire.
- Keep your weapon on SAFE until you are ready to fire.

X. Techniques are applied with slow, steady pressure to the point where the aggressor is uncomfortable; the aggressor must then tap out. Marines must immediately release pressure or stop the technique. The aggressor will tap out by tapping on the deck three times, on himself three times, on the Marine three times, or by verbally saying "tap-tap-tap."

XI. Never hold a choke for more than 5 seconds in training. The aggressor should never become light-headed during a choke.

XII. Do not apply pressure to the aggressor's trachea during training.

XIII. Students being thrown should execute proper break-falls.

XIV. Conduct all practical application periods utilizing approved training gear.

XV. Ensure that calf-on-calf contact is being made.

For Official Use Only

This Page Intentionally Left Blank.

Glossary

LINE	linear infighting neural-override engagement
MAIT	martial arts instructor trainer
MCMAP	Marine Corps Martial Arts Program
MCRP	Marine Corps reference publication
MOS	military occupational specialty
PME	professional military education

THIS PAGE INTENTIONALLY LEFT BLANK.

References

Marine Corps Orders (MCOs)

1500.59   Marine Corps Martial Arts Program (MCMAP)

P3500.72A   Marine Corps Ground Training and Readiness (T & R) Program

Marine Corps Institute (MCI)

0337   Leading Marines

THIS PAGE INTENTIONALLY LEFT BLANK.

For Official Use Only

Printed in Poland
by Amazon Fulfillment
Poland Sp. z o.o., Wrocław